Extranuclear Genetics

Extranuclear Genetics

Geoffrey Beale, F.R.S.
Royal Society Research Professor at Edinburgh University

and

Jonathan Knowles
Royal Society Research Fellow at Centre de Génétique
Moléculaire, Gif-sur-Yvette, France

Edward Arnold

First published 1978
by Edward Arnold (Publishers) Limited
25 Hill Street, London W1X 8LL

ISBN: 0 7131 2641 8

Filmset by Keyspools Ltd, Golborne, Lancs
Printed in Great Britain by
Fletcher & Son Ltd, Norwich

Preface

At the present time considerable effort is being devoted to the study of the genetic properties of mitochondria and chloroplasts. These cellular organelles have only recently been conclusively shown to contain their own genetic systems, distinct from the nuclear system and based on strands of DNA which show more resemblances to the DNA of bacteria and viruses than to the chromosomes of higher animals and plants. This has led to a revival of interest in the old idea that cellular organelles have evolved from free-living microorganisms, by way of symbiosis.

We have approached this subject by considering the whole range of examples of extranuclear heredity, not only in higher organisms but also in bacteria. We have organized the material into chapters dealing with mitochondria, chloroplasts, endosymbionts and plasmids. This leaves a large residue of unclassified phenomena, usually designated 'cytoplasmic heredity', whose material basis is at present not known. It may be that much of this residue will eventually be accommodated in one or other of the categories which we have specified. If so, nearly all types of extranuclear genetic phenomena will be assignable to some DNA-containing particle, giving a coherent theoretical basis to the subject; though as we show in this book, there is a good deal of heterogeneity, or 'untidiness', in the different examples. By contrast, the theory of nuclear or classical genetics is remarkably consistent throughout the biological world.

Organelle genetics, and even more the genetics of bacterial plasmids, are such rapidly advancing areas of research that we cannot hope our account is completely up-to-date even at the time of writing; and obviously in a small book we cannot give a thorough review of the relevant literature. We have tried to support our statements with appropriate references to original publications, but wish to stress that we have made no attempt to specify the first publication of particular pieces of research. We feel that it is more useful in a general book of this type for the reader to be informed of the most recent accounts, and if possible of readily available review articles, rather than to be told the name of the scientist who claims priority for a particular discovery. Readers are assumed to have an elementary knowledge of genetics and molecular biology, but we have striven to make the account intelligible to a wide audience.

It will be noticed that in some places we have not hesitated to state opinions which may not be those of some of the workers who have done the original research. We hope that this makes for a more readable account than would result from an uncritical restatement of a mass of inconclusive data, even though it will probably turn out that some of our interpretations are incorrect. Some topics on which a large amount of research has been done are treated relatively briefly. This applies particularly to the recent work on *Chlamydomonas*, which has been extensively discussed in the publications of R. Sager and N. W. Gillham, and also to the abundant German literature on extranuclear genetics of higher plants. We have been much aided by the comments of a number of colleagues, and would especially thank A. Adoutte, C. Curtis, B. Dujon, R. J. Ellis, N. Gillham, A. Kingsman, G. Ledoigt, A. Linnane, J. R. Preer, R. Sager and A. Tait, none of whom should of course be held responsible for any of our errors. In addition we wish to acknowledge with thanks permission to reproduce a number of illustrations which have appeared elsewhere, and are also very grateful for unpublished photographs which have been presented to us, as indicated in the text.

We should of course appreciate any comments pointing out errors of fact or interpretation.

G.B.
J.K.

Edinburgh, 1977

Contents

1 Introduction

Genetics is firmly based on the belief that genes are located on chromosomes in the cell nucleus. This is true of higher animals and plants with which the main theory of genetics was developed. It is also true of bacteria, which may lack the complex nuclei and chromosomes of higher organisms but nevertheless contain a strand of DNA on which most genes are situated. The theory of genetics is remarkably consistent and applicable to all forms of life, from viruses to mammals.

However, for many years there have been biologists—and others, who have been unwilling to accept this notion of a nuclear monopoly and have sought to show that other regions of the cell are also concerned with heredity. Such feelings may have been based, to some extent, on a healthy reaction against the extremely rigid, mechanistic concepts of classical geneticists, with their notions of genes as beads on strings, breaking and rejoining, and splitting into two, while the rest of the cell was considered as a kind of tank into which the gene products were discharged. Thus Morgan (1926) wrote 'The cytoplasm may be ignored genetically'.

From the earliest days of genetics, just after the rediscovery of Mendel's principles in 1900, the doubters have not been without some support for their doubts. In 1909, it was already known that in plants some variations did not obey Mendel's laws, and the number of such exceptions has steadily grown since that time. When crosses were made between different species of plants, many characters seemed to show non-Mendelian inheritance, and were transmitted more readily through female than male reproductive cells. In fact some people even held the extreme view that Mendelism only applied to relatively 'trivial' characters, such as whether seeds were round or wrinkled, while important characters, constituting the raw material of evolution, were inherited in some other way. All this, of course, seems absurd to us now that we are familiar with the elaborate maps of genes for almost every conceivable property of *Drosophila melanogaster* and *Escherichia coli*.

The number of non-Mendelian phenomena continues to grow, however. To mention a few well-known examples, apart from the chloroplast variants of plants, there are the 'killer' paramecia, the CO_2-sensitive drosophilae, the 'petite' yeasts, and many cases of male-sterile plants, all controlled by extranuclear genetic factors. Admittedly the number of such examples is small by comparison with the number controlled by nuclear genes, and some of the extranuclear phenomena may seem to be of a mildly pathological nature. This cannot be said however of all cytoplasmic constituents, since mitochondria, which occur in nearly all forms of life more complex than bacteria, and chloroplasts, which are essential to virtually all plants, are now known to contain their own separate genetic systems.

The notion of extranuclear heredity gained respectability when it was discovered that chloroplasts and mitochondria contain DNA, chemically distinct from the DNA in the nucleus. Kappa particles in killer paramecia were also found to contain DNA, and attempts were made to show that other cytoplasmic structures, such as the ciliary basal bodies or kinetosomes contained it (though these latter now seem not to do so). The finding of DNA in various bodies outside the nucleus made it possible to consider cytoplasmic heredity as really a kind of accessory 'nuclear' heredity, due to the presence of fragments of nuclear material in the cytoplasm, possibly in the form of symbionts or viruses. Thus all genetics could happily still be considered to·be based on 'nuclear' structures, or something like them.

This line of thinking need not be pursued any further. Let us admit that there are important genetic phenomena controlled by elements outside the main mass of DNA in the nucleus and these 'extranuclear' elements, as we shall from now on call them, do not behave in precisely the same way as those within the nuclei. The extranuclear elements do not pass through cycles of mitosis and meiosis, and the various characters controlled by such elements do not exhibit precise Mendelian behaviour. In this book we plan to describe the more clearly analysed cases of extranuclear heredity and we shall classify our material as far as possible into groups based on particular DNA-containing extranuclear structures or organelles. Due to incomplete knowledge, this is not always possible and there is a large residue of examples whose material basis is unknown. Eventually it may turn out that much of this residue can also be accommodated in one or other of the already specified types. The extranuclear factors with a known material basis can be classified as follows: (1) those based on the mitochondria and chloroplasts, cell organelles of very wide occurrence; (2) those based on structures of more restricted occurrence, some being analogous to symbionts, viruses, etc. and (3) the separate category of bacterial plasmids.

These diverse examples occur in a wide variety of organisms—bacteria, fungi, protozoa, algae, angiosperms and higher animals. The methods used to study extranuclear factors are also diverse, some being based on the conventional hybridization techniques of genetics, others on the more modern methods of molecular biology. Ideally a given example should be studied by a variety of techniques, but that is seldom done. Consequently, evidence for one example derived by one technique is sometimes combined with evidence for another example using a different technique, whereby a general theoretical structure which may be insecure is developed, as will be discussed later.

The principal methods which have been used in the study of extranuclear genetics are summarized in the following paragraphs.

Reciprocal hybridization If one has two different types of organisms, A and B, they may be crossed by using A as the female parent and B the male (A♀ × B♂), or vice versa (B♀ × A♂). If the progeny from such reciprocal crosses are different, that is presumptive evidence for an extranuclear genetic effect, since both male and female gametes as a rule supply equal quantities of chromosomal material but the female parent usually contributes much more cytoplasm than the male. Differences in the progeny of reciprocal crosses may, however, be due to causes other than the presence of extranuclear factors. One such cause is the so-called 'maternal effect', a number of examples of which have been described (see Whitehouse, 1968). Perhaps the best known is that concerned with the direction of coiling of the shell of the snail *Limnaea*. The direction is controlled by nuclear genes, but the expression of these genes is delayed, the coiling genes of the mother being expressed in the progeny, due, presumably, to their action being on the orientation of certain constituents of the egg cytoplasm. True extranuclear inheritance—that is, inheritance based on genes in the cytoplasm—can be distinguished from maternal effects by making a series of repeated backcrosses of the type

$$A♀ \times B♂$$
$$\times B♂$$
$$\times B♂ \ldots,$$

whereby the nuclear genes of B are eventually combined with the cytoplasmic genes of A, if there are any. If the progeny obtained after a series of, say ten, backcrosses—by which stage virtually all nuclear genes of A will have been replaced by those of B—are different from B, it is concluded that there are some extranuclear genes in A. A number of examples in which this method has been used will be described later (see pp. 21, 88, 108). The method can sometimes be used even when the two types of gamete are morphologically identical, e.g. in *Chlamydomonas*, where cells of one mating type transfer certain extranuclear (chloroplast) factors to the zygote, and cells of the other mating type usually do not, even though the zygote is formed by a total fusion of the two gametes. The method can also be used in *Paramecium*, where conjugation produces a reciprocal exchange of male and female gamete nuclei from each conjugant, while little cytoplasm is exchanged (see p. 86). On the other hand, in some plants, such as *Pelargonium*, the pollen tubes contribute as much of some cytoplasmic factors to the next generation, as does the egg, and the method is of little use.

Heterokaryon formation In fungi the technique of heterokaryon formation is useful for the study of extranuclear genetics, and can be used when sexual processes are not available (Jinks, 1964). Two strains are brought together and the hyphae allowed to anastomose, producing heterokaryons—hyphae with two different kinds of nuclei—as well as mixtures of cytoplasm in some cases. By various techniques, e.g. by making cultures from uninucleate spores, one can re-establish homokaryons, the nuclei of which can be identified if the original strains contained different nuclear genes with characteristic phenotypic effects. If now some characters originally in one strain, show recombination with the *genic* markers of the other strain, that is evidence for the presence of extranuclear genetic factors. Examples of this method are given on pp. 21, 99, 105, 106. A modified version of the method has also been used in *Amoeba* where heterokaryons can be made by injecting nuclei artificially from one cell to another (see p. 113), and subsequently obtaining homokaryons by cutting the cells into two or allowing them to divide into uninucleate cells.

Microinjection techniques Cytoplasmic components, e.g. mitochondria, can be injected into different cells and new nuclear–cytoplasmic combinations established. Such techniques have been used successfully in *Neurospora* (p. 21), *Paramecium* (p. 23) and *Amoeba* (p. 113).

Non-Mendelian behaviour Apart from the above methods, by which different kinds of cytoplasmic and nuclear factors are deliberately combined, information indicating the existence of extranuclear genes can sometimes be obtained from the observation of non-Mendelian behaviour, since segregation and recombination of cytoplasmic genes may take place not only at meiosis, but also during mitotic divisions. Such observations are of use in an organism like yeast, where sexual fusion results in the fusion or admixture of all the cell contents, both nuclear and cytoplasmic. Moreover, the absence of linkage with known chromosomally located genes can also be used as an indication of extranuclear inheritance in organisms like *Drosophila* or maize, where all the chromosomes are well marked with genes.

Infection In some cases there may be a rapid spread of a condition from one strain of an organism brought into contact with another under circumstances where nuclear passage does not occur. A number of examples of this type of behaviour are known in fungi (p. 104), and it is possible, according to the views of some workers, to obtain transmission of some infectious material across the graft boundaries in plants (p. 110). Moreover, in some cases infection can be brought about artificially by injection or implantation, as in *Drosophila* (pp. 89, 92). Of course, the nature of the infectious agent, if known, can immediately establish its extranuclear nature.

Identification of the nature of extranuclear genetic factors While the existence of extranuclear factors, if present, is relatively easily demonstrated, to establish their precise nature and location in the cell is sometimes much more difficult. In some cases, where a conspicuous symbiont is present in cells displaying a given effect and absent in cells not showing the effect, the agent can be easily identified by microscopical examination, as in the case of the *Paramecium* kappa particles (p. 81), or the spiroplasma-like particles in some *Drosophila* species (p. 90). In other cases, electron microscope studies may be necessary, as with the virus-like sigma factor of *Drosophila* (p. 95). But in still other cases, complex studies are required to establish the exact location of the extranuclear factor concerned, as with mitochondrial factors in yeast. For some extranuclear factors there has been long controversy as to their exact location. This applies particularly to factors affecting chloroplast characters in higher plants, and to some extent even in *Chlamydomonas*, as will be discussed later (pp. 55, 60).

By one or other of the above described methods, it has been established that there are genetic elements associated with various extranuclear structures, which (with rare exceptions, see p. 97), contain their own specific DNA. All of these structures are also to a greater or lesser degree controlled by genetic factors in the nucleus, and this leads us to ask: what proportion of the characters of these extranuclear structures are controlled by nuclear DNA and what proportion by extranuclear DNA, and how do the two types of genetic system interact? To answer these questions, various techniques of genetics and molecular biology have been applied, as will be described in detail in later chapters. The amount of information of this sort, available for different extranuclear systems, is very uneven, but in all cases complex interactions occur, and this leads to the final question: how have these extranuclear systems evolved? This will be discussed, though not of course answered, in the final chapter.

At many points in the account we shall have occasion to compare and contrast two kinds of organism denoted prokaryotic and eukaryotic, terms used by Chatton (1925), and Dougherty (1957) and discussed in more detail by Stanier and van Niel (1962). It may be useful to expand a little on this distinction here. On the one hand the prokaryotes, which include bacteria, blue-green algae and viruses, contain DNA lacking histone (or in the case of some viruses, RNA in place of DNA); they do not have a membrane around the nuclear region, or indeed around any other internal constituent; they do not undergo mitosis to ensure the regular distribution of the nuclear material to daughter cells by means of spindles, centromeres, etc., such distribution being accomplished in some other way; and finally, they do not undergo regular sexual processes, so have no need of meiosis, though various mechanisms for bringing about genetic recombination are known, some of which will be described later (p. 71). On the other hand, the eukaryotes, ranging from protozoa, algae and fungi to all the more complex multicellular organisms, contain distinct nuclei bounded by a membrane and containing true chromosomes which have an elaborate structure and are capable of undergoing mitosis and meiosis. In eukaryotes there is also a distinct part of the cell, the cytoplasm, separate from the nucleus, and containing a number of membrane-bound structures, notably the mitochondria, concerned with respiration, and (in plants) the chloroplasts, concerned with photosynthesis. Since these cytoplasmic membrane-bound structures or organelles show a number of resemblances to prokaryotic organisms, as will be discussed in later chapters, we may define a eukaryotic cell as one containing not only a highly developed nucleus, but also some membrane-bound prokaryotic structures outside the nucleus. Further discussion on the differences between prokaryotes and eukaryotes is found in the volume edited by Charles and Knight (1970).

Thus, study of extranuclear heredity leads us into questions of general biological interest, such as the evolution of the cells characteristic of higher organisms. It also reveals, as will be seen especially in Chapter 4, that the biological world is, so to speak, littered with an assortment of particles, structures and pieces of DNA and RNA, which do not fit into a neat pattern or have an obvious place in a scheme for the evolution of ever more perfect higher forms of life.

2 Mitochondria

Introduction

Mitochondria are very important constitutents of the cytoplasm and occur in the overwhelming majority of eukaryotic cells. The basic respiratory processes, leading to the production of energy in the form of ATP, take place in mitochondria, which are essential for eukaryotic life as it exists now. Damage to mitochondria by drugs or other agents usually results in death, except in some yeasts and a few other organisms which have an alternative anaerobic pathway for energy metabolism.

However, in this book we are primarily concerned not with the metabolic, energy-producing, functions of mitochondria, but with their genetic properties. It has been shown that mitochondria contain a small amount of DNA, roughly equivalent in size to that of bacterial viruses (Nass & Nass, 1963), and this mitochondrial DNA is qualitatively distinct from that in the nucleus. More recently, as suggested earlier, though without rigorous proof, mitochondria have been shown to contain their own genes, controlling a small but essential fraction of the mitochondrial characteristics. Our aim is to describe this genetic component of mitochondria. This is now the object of a great deal of intensive research which is proceeding so rapidly that it is impossible to present a completely up-to-date account.

The genetic approach to the study of mitochondria, quite apart from its intrinsic interest to geneticists, has proved remarkably productive of new ideas. It has stimulated research workers to dissect the structures and metabolic processes into their many component parts, each having their individual roles but contributing to the functioning of the whole, in a way which would have been difficult or impossible by morphological or biochemical studies alone. Genetic knowledge also makes possible an interesting (though inconclusive) discussion of the evolutionary history of mitochondria. For many years there has been speculation about their homologies and origins. As early as 1890, it was suggested by Altmann that mitochondria might be derived from bacteria, and recently this endo-symbiotic theory has been revived (Margulis, 1970). However, other quite different views have been expressed (Raff & Mahler, 1975), and there is no general agreement about the matter. We will discuss this later (pp. 116–121).

The hereditary elements of mitochondria, though very minor by comparison with the genes in the nucleus, are of unusual interest. Indeed in some respects the smallness of the mitochondrial genome is of particular value, as it makes feasible certain types of investigation. This applies to problems like the mechanism of replication of DNA in organisms from which the DNA can be isolated intact, and the possibility of mapping the complete mitochondrial gene sequence in certain organisms. It is also possible that mitochondria, like some plasmids, may reveal information about genetic systems that is not elsewhere accessible.

The structure and properties of mitochondria

Introduction Mitochondria vary enormously in size, shape and structural content in different organisms, in different tissues and in cells growing under different conditions. Figure 2.1 shows the mitochondria in *Paramecium* as seen by phase microscopy. In Figures 2.2–2.5 and 2.12 we show electron microscope pictures of a few types of mitochondria. Many more are illustrated by Munn (1974). All mitochondria are bounded by

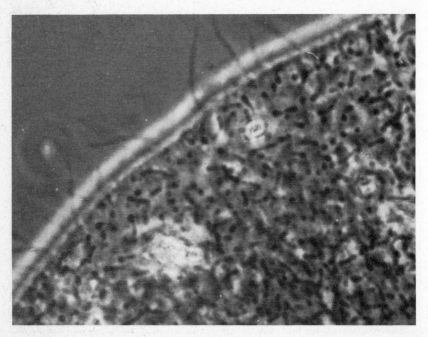

Fig. 2.1 Mitochondria of *Paramecium* by phase microscopy. (By kind permission of R. Perasso & J. Beisson.) × 2000.

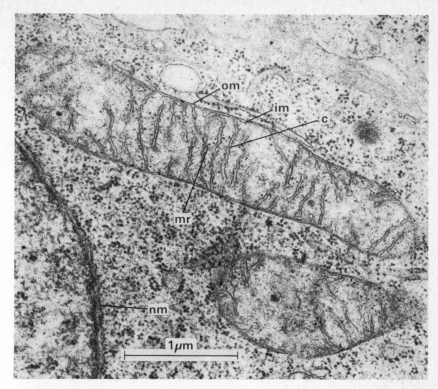

Fig. 2.2 Electron micrograph of mitochondria in the pancreas of chicken embryo. The outer membrane (om), inner membrane (im), crista (c) and mitochondrial ribosomes (mr) are all clearly visible. A part of the nucleus surrounded by nuclear membrane (nm) is also seen. (By kind permission of J. André.)

outer and inner membranes within which is the matrix containing a vast array of soluble enzymes and metabolic intermediates, as well as DNA and ribosomes. In the following sections a few details which are particularly relevant to the subject matter of this book will be mentioned.

Inner membrane and cristae (see Fig. 2.6) These are perhaps the most characteristic structural features of mitochondria, in so far as their appearance in the electron microscope is concerned. The inner membrane is extremely complex and is the site of many important biochemical activities connected with electron transport, oxidative phosphorylation and protein synthesis. Usually the inner membrane bears an elaborate arrangement of lamellar or tubular cristae. In yeast, under some conditions, such as a lack of oxygen or presence of catabolite repressors (e.g. glucose), the cristae are much reduced or even lacking entirely, but on restoration of favourable conditions, they develop normally again. Moreover, in organisms with complex life cycles, cristae may be very rudimentary at one stage (e.g. the blood forms of mammalian malaria parasites), but fully developed at other stages (e.g. the insect forms of malaria parasites).

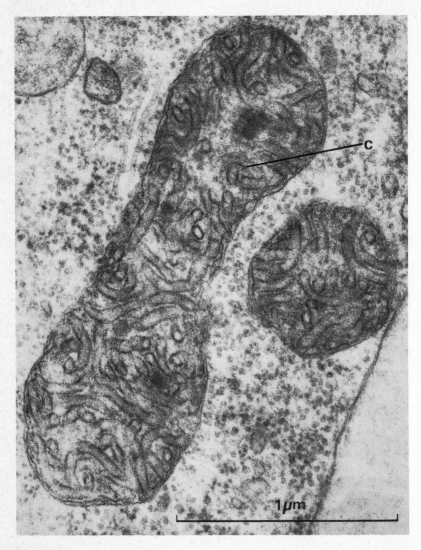

Fig. 2.3 Electron micrograph of mitochondria in *Paramecium aurelia*, one in cross section and another in longitudinal section. The inner membrane is continuous with the tubular cristae (c) which are typical of many ciliates. (With permission from A. Jurand and G. Selman, 1969. *The Anatomy of Paramecium aurelia*, Macmillan, London, 139.)

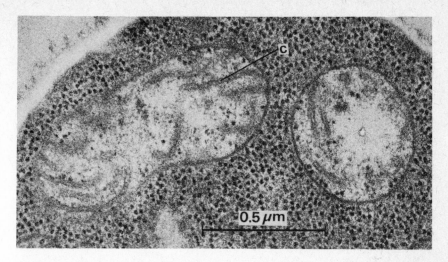

Fig. 2.4 Electron micrograph of mitochondria in a haploid yeast cell, *Saccharomyces cerevisiae*, in mid exponential phase grown on glycerol. The form of the mitochondria is very variable depending on the exact state of growth. Several cristae (c) are seen in each profile. (By kind permission of B. Stevens.)

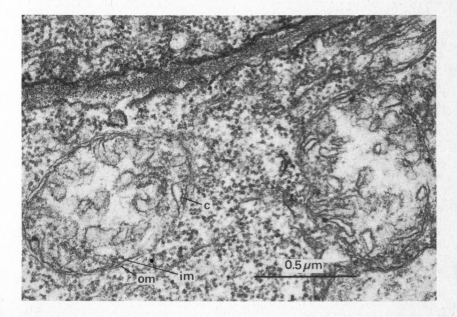

Fig. 2.5 Electron micrograph of mitochondria of a pear tree (*Prunus*). The inner (im) and outer (om) membranes and the cristae (c) are visible. (By kind permission of J. André.)

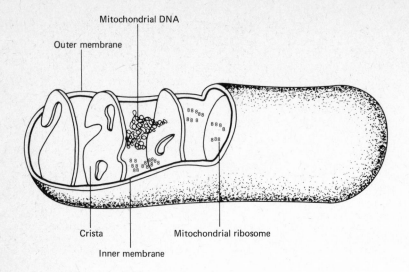

Mitochondrial DNA

Outer membrane

Crista

Inner membrane

Mitochondrial ribosome

Fig. 2.6 Mitochondrial structure. A part of the inner and outer membrane has been removed to show the main structural features of a 'generalized' mitochondrion. The DNA is shown as a 'nucleoid' attached to a crista. The ribosomes are also often attached to cristae and may sometimes be seen as polysomes.

Ribosomes In favourable preparations, ribosomes can be seen by electron microscopy, attached to the mitochondrial inner membrane or within the matrix. Sometimes they can be seen to be collected together in polysome-like chains (Fig. 2.7) (Vignais *et al.*, 1969, in *Candida*; Avandhami & Buetow, 1972, in *Euglena*). Mitochondrial ribosomes have been extracted and purified from several types of organism. In general these ribosomes sediment more slowly (i.e. they are smaller) than cytoplasmic ribosomes, but there is considerable variation between different groups of organisms. According to Dawid (1972), the mitochondrial ribosomes fall into two main groups, one comprising those in various fungi and in *Euglena*, with an S value of 70–74; the other including various higher animal species, with an S value of 55–60. To these should now be added a third group for the ciliates, and possibily also the higher plants, in which the mitochondrial ribosomes have an S value of 80, i.e. one close to that of the cytoplasmic ribosomes. Some data on this subject are summarized in Table 2.1. Formerly it was thought that the S values of mitochondrial ribosomes resembled those of bacteria, but this is obviously not a general rule.

However, in some respects, such as sensitivity to certain drugs, mitochondrial ribosomes do resemble those of bacteria. Protein synthesis on bacterial and mitochondrial ribosomes is inhibited by chloramphenicol (CAP) and erythromycin (ERY), but not by cycloheximide; while protein synthesis on cytoplasmic ribosomes is inhibited by cycloheximide, but not by CAP or ERY.

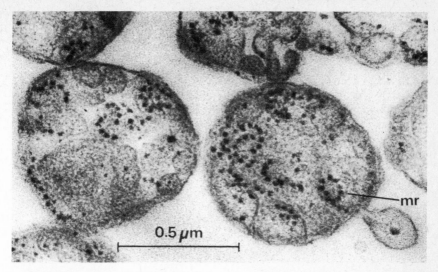

Fig. 2.7 Electron micrograph of a section through a pellet of isolated mitochrondria from the yeast *Candida utilis*. This preparation has been stained to show clearly the mitochondrial ribosomes (mr) which often appear grouped as polysomes. (With permission from Vignais *et al.* (1969) and North Holland, Amsterdam.)

In general, as pointed out by Dawid (1972), mitochondrial ribosomal RNA shows no sequence homology with cytoplasmic ribosomal RNA from the same species.

Transfer RNA It is noteworthy that mitochondria contain their own sets of t-RNAs, which are distinct from the cytoplasmic t-RNAs in various physical characteristics. The number of mitochondrial t-RNAs so far identified is 25 in yeast (*Saccharomyces cerevisiae*). In *Neurospora* even more mitochondrial t-RNAs have been reported (Blossey & Küntzel, 1972). (See further discussion on p. 34).

Mitochondrial DNA and its replication As already mentioned, mitochondria contain a small amount of DNA. Under some conditions this can be seen by electron microscopy as a collection of fine filaments in the matrix (Nass & Nass, 1963; and see Fig. 2.8), but much more detail can be made out when the DNA is released from the mitochondria by osmotic shock followed by spreading over electron microscope grids by the Kleinschmidt technique. The DNA is then seen to consist of covalently closed circles, which have a length of about 5 μm in higher animals, 20 μm in *Neurospora*, 25 μm in *Saccharomyces* and 30 μm in peas (Borst, 1972; Kolodner & Tewari, 1972; Ryan *et al.*, 1974). Ciliate protozoa seem to be exceptional in that in the two examples so far studied, only linear filaments have been seen, their lengths being 15 μm in *Tetrahymena pyriformis* (Borst,

Table 2.1 Sedimentation values of mitochondrial ribosomes and subunits.

	Organism	Monomer	Sub-units		RNA	
			Large	Small	Heavy	Light
Fungi	*Neurospora* *Saccharomyces* *Candida*	70–74S	50S	40S	21–24S	14–16S
Higher animals	Rat *Xenopus* HeLa cells	55–60S	40S	33S	16–17S	12–13S
Ciliates	*Tetrahymena* *Paramecium*	80S	55S	55S	21S	14S
Algae	*Euglena*	71S	50S	32S		
Higher plants		80S				
(Cytoplasmic ribosomes of eukaryotes (yeast, HeLa))		80S	60S	40S	26–28S	18S
(Bacterial ribosomes *E. coli*)		70S	50S	30S	23S	16S

Notes: (1) Data from Dawid (1972), except for *Paramecium* (Tait & Knowles, 1977) and for *E. coli* (Nomura *et al.*, 1974). (2) The two subunits of *Paramecium* mitochondrial ribosomes, though indistinguishable by S values, can be distinguished by electron microscope observations of the monomers. (3) These values are approximate only, since considerable variations occur with different techniques of measurement.

1972) and 14µm in *Paramecium aurelia* (Cummings *et al.*, 1976), though this does not necessarily imply that circular mitochondrial DNA does not occur in living cells of these ciliates.

There is no obvious resemblance between the base composition of mitochondrial and nuclear DNA of the same species: the GC content of mitochondrial DNA may be higher, the same, or lower than that of nuclear DNA. No nucleotide sequences common to nuclear and mitochondrial DNA can be detected by DNA/DNA hybridization studies.

Like other small circular DNAs (e.g. those of plasmids), mitochondrial DNA shows superhelical turns or supertwists.

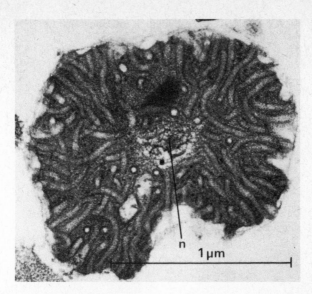

Fig. 2.8 Electron micrograph of an isolated mitochondrion of *Tetrahymena* showing the fibrillar nucleoid (n). By kind permission of R. Charret.

Recently, restriction endonucleases have been used in the study of mitochondrial DNA. Such enzymes break the DNA strands at specific sites, depending on the presence of particular sequences of bases. Different species of mammal show different restriction enzyme patterns (Potter *et al.*, 1975), as is also found in the mitochondrial DNA of different species (syngens) of *Paramecium aurelia* (Cummings *et al.*, 1976).

The replication of mitochondrial DNA is thought to take place by a 'modified Cairns mechanism' (Borst, 1972; Robberson *et al.*, 1972). A suggested scheme is illustrated in Fig. 2.9. (See also Fig. 2.10.)

From electron microscope studies of mammalian mitochondrial DNA, it has been concluded that replication starts at a particular site on the light (L) strand, and the 'parental' heavy (H) strand is displaced, producing the displacement or D loop. Replication of the parental L strand continues as the D loop increases in size. After a delay, the displaced parental H strand starts to replicate. There is thus an 'asymmetry' in the replication of the two strands—i.e. one replicates after the other. According to Kasamatsu and Vinograd (1974) replication is unidirectional in each strand.

The replication process appears to be somewhat different in the two ciliates *Tetrahymena* and *Paramecium*. In *Tetrahymena* replication starts at the midpoint of the linear DNA molecule and proceeds simultaneously in both directions, so that the replicating intermediate is an 'eye' type of structure (Upholt & Borst, 1974). By contrast, in *Paramecium* replication is unidirectional and in this case starts at one end, the intermediate being in the form of a 'lariat'. At the completion of replication a linear dimer is formed,

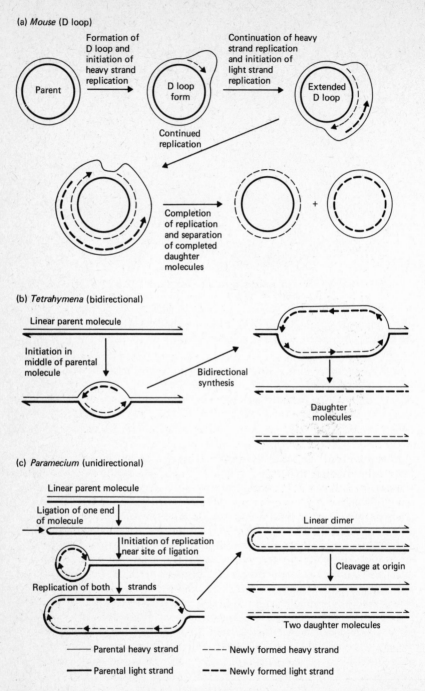

Fig. 2.9 Models of the replication of different mitochondrial DNAs. (Redrawn with permission from Robberson *et al.* (1972), Upholt and Borst (1974) and Goddard and Cummings (1975).

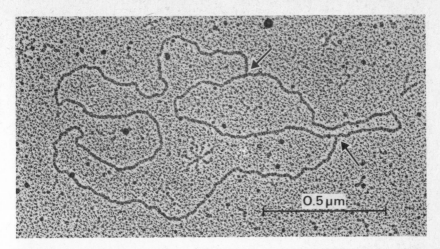

Fig. 2.10 Electron micrograph of a molecule of mitochondrial DNA isolated from rat liver cells. This molecule shows two replication forks (arrowed). (With permission from D. Wolstenholme, K. Koike and P. Cochran-Fouts (1973), Chromosome structure and function, *Cold Spring Harbor Symp.*, **38**, 267–80.)

twice the length of the normal linear strand. Eventually the dimer breaks at the junction of the two monomers (Goddard & Cummings, 1975).

Replication and fusion of mitochondria   Mitochondria in living cells, when observed by phase contrast microscopy have been observed to 'divide', but since they are near to the limits of resolution by light microscopy, detailed visual observations of the process *in vivo* are impossible. Fixed preparations examined by electron microscopy may show apparently dividing structures (Fig. 2.11), i.e. two roundish objects joined by a narrow isthmus. This does not prove anything in view of the constant changes in shape, and even fusions, which mitochondria undergo.

In spite of uncertainty about the details, it is now generally accepted that mitochondria do not arise *de novo*, nor by some invagination of membranous structures as suggested by Bell and Mühlethaler (1964), but by replication of the pre-existing mitochondria or at least of some part of them. Certainly this statement must apply to the mitochondrial DNA, though in some organisms, as already mentioned (p. 10), there are stages containing only rudimentary mitochondria, out of which fully-developed forms arise at later stages. It must be assumed that in rudimentary forms the mitochondrial DNA is retained intact. What other mitochondrial components are thus obligatorily retained is unknown.

As regards 'fusion' of mitochondria, genetic evidence will be presented later showing that recombination of mitochondrial genes occurs, at least in some organisms, such as the yeast *Saccharomyces cerevisiae*. This presumably involves a fusion of mitochondria, or passage of DNA from one

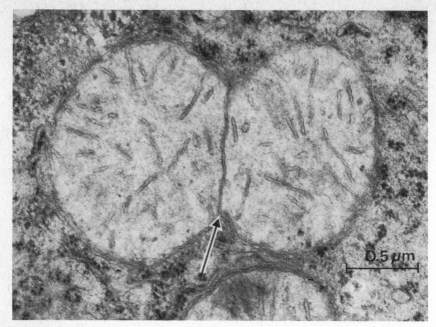

Fig. 2.11 Electron micrograph of a mitochondrion in a mouse liver cell. The presence of central dividing membranes (arrowed) suggests that this organelle is in the process of dividing. (By kind permission of T. Cavalier-Smith.)

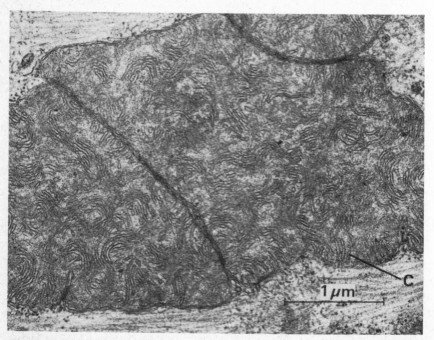

Fig. 2.12 Electron micrograph of the mitochondria of a flight muscle of *Locusta migrata*. The extensive system of cristae (c) is evident. (By kind permission of J. André.)

to another. Light microscope observations indicate that fusions may occur, but the details are not clear. In yeast, evidence based on electron microscope studies of serial sections indicates that under certain conditions the mitochondria in a cell may fuse into a single reticulate structure (Hoffman & Avers, 1973; Stevens, 1977).

Principal phenotypic effects of mitochondrial genes

Introduction In this section the genetic evidence for the inheritance of certain mitochondrial characters will be described, with special attention to those aspects controlled by mitochondrial genes. Here we will deal only with the easily observable consequences of mitochondrial genetic variation, and leave a more detailed analysis of the underlying biochemical stages till later (pp. 31–7).

The characters to be considered are classified mainly in two groups: (1) those concerned with deficiencies in the respiratory system, and (2) those concerned with drug resistance. The two groups are not mutually exclusive, since some drug-resistant mutants affect components of the respiratory system. Moreover, certain other characters cannot be included in either group and will be considered separately (p. 38). This arbitrary classification is made for convenience of presenting the material.

Only a few types of organism have so far been extensively used in studies of mitochondrial variations by traditional genetic methods (i.e. hybridization and analysis of the progeny): three fungi and one protozoan— *Saccharomyces cerevisiae, Neurospora crassa, Aspergillus nidulans* and *Paramecium aurelia.* It is likely that this restriction in the range of organisms studied may have led to a bias in the kind of information obtained, but this can only be shown when more organisms have been examined. Biochemical work, by contrast, has extended over the whole range of eukaryotic organisms. This preponderance of biochemical over genetic studies may sometimes have resulted in the drawing of unwarranted general conclusions regarding genetic mechanisms.

The 'petite' yeasts The production of 'petite' colonies in cultures of baker's yeast, *Saccharomyces cerevisiae,* was first described in 1949 by Ephrussi and collaborators (Ephrussi, 1953). These strains grow slowly, produce small colonies on agar plates containing little glucose and have a very low level of respiration, but are able to grow due to their ability to obtain energy by use of the alternative fermentation pathway. They arise spontaneously at a rate (about 1/500 per cell generation) much higher than that of most genic mutations, and by treatment with ultraviolet light or some chemicals, such as acriflavine or ethidium bromide, the rate can be raised almost to 100%.

Several types of petite mutant, differing in their mode of inheritance, are known. They were classified as follows: (1) neutral (ρ^0), (2) suppressive

(ρ^-), and (3) segregational. Classes (1) and (2) were shown to be due to loss or modification of a cytoplasmic genetic factor, denoted ρ (or rho), while class (3) was due to nuclear gene mutations. In brief, the evidence for this is as follows. Class (3) petites show Mendelian inheritance, i.e. when normal and segregational petites are crossed, the diploid hybrids are normal, and after sporulation, groups of four ascospores arise showing a segregation of 2 normal:2 petite. Class (1) (neutral) petites, on the other hand, when crossed with normals, yield only normals in all subsequent generations. Class (2), the suppressive petites, when crossed with normals, give more complex results. These crosses result in the formation of varying proportions of normal and petite progeny. When a highly suppressive petite stock is used in a cross with wild type all the progeny are petite.

These results were interpreted as showing that the neutral and suppressive petites are abnormal in regard to a cytoplasmic factor (ρ), which is present in normal cells and required for respiration. The segregational petites were assumed to contain the normal ρ factor, but respiration was defective here because of the presence of one of a number of recessive nuclear genes (now denoted *pet-1-51* (Plischke *et al.*, 1975)).

Evidence was obtained early in this work implicating mitochondrial variation in the cytoplasmic petites (Ephrussi, 1953). These petites lack various mitochondrial components, such as cytochromes $a + a_3$ and b, though cytochrome c is present unaltered. In addition, the mitochondria of petites show some morphological abnormalities. In light microscope preparations, certain staining reactions (e.g. Nadi) typical of normal mitochondria are negative, and electron microscope studies show that under certain conditions the inner membrane structures of mitochondria in cytoplasmic petites are poorly developed (Yotsuyanagi, 1962a & b). None of this is proof that the genetic determination of cytoplasmic petites lies within mitochondria, though no indications have been found of any other cytoplasmic location of the genetic factors concerned.

The evidence for a mitochondrial location of these factors became clearer after the discovery that the mitochondrial DNA in cytoplasmic petites is different from that in normal yeast cells (Mounolou *et al.*, 1966). Neutral petites are now known to have lost all their mitochondrial DNA, while suppressive petites have incomplete mitochondrial genomes containing many deletions and duplications (Nagley & Linnane, 1972; Faye *et al.*, 1973). The total amount of DNA in suppressive petites is often not much smaller than in normals because of the phenomenon of amplification of DNA in petites, whereby any part of the genome may become duplicated many times and rearranged. Even when the amount of mitochondrial DNA in cytoplasmic petites is similar to normal, the extensive rearrangements often cause qualitative differences (Hall *et al.*, 1976). Clearly there is an association between mitochondrial DNA variation and mutation to cytoplasmic petite, and in view of the known genetic function of DNA, it is not unreasonable to accept that the change in mitochondrial DNA is the cause and the petite phenotype is the effect.

Apart from the petites, many defects in the respiratory system of yeast have now been shown to be caused by mutations involving small segments of DNA or, point mutations, generally denoted *mit-* (Tzagoloff *et al.*, 1975). These will be referred to in a later section (p. 37).

Respiratory variants in *Neurospora crassa* As already stated *Saccharomyces cerevisiae* is able to survive when respiration is at a very low level and hence can live with defective mitochondria or in the absence of oxygen, though at the expense of a slow growth rate. By contrast *Neurospora* cannot live under anaerobic conditions and can only tolerate partial impairment of the respiratory system (Bertrand & Pittenger, 1972a). Nevertheless, a number of cytoplasmically inherited variants showing such partial impairments are known in this fungus.

In 1952, Mitchell and Mitchell described the type known as 'poky' (later denoted *mi-1*), which grows slowly, and respires poorly, being defective in various cytochromes and respiratory enzymes. Subsequently many other similar variants were described (Srb, 1966; Fincham & Day, 1971; Bertrand & Pittenger, 1972b; Bertrand *et al.*, 1976) most of which arose spontaneously, though some, such as the *SG* types described by Srb (1963), appeared following treatment with acriflavine.

Inheritance of poky, and other similar variants of *Neurospora*, was found to be strictly 'maternal', i.e. if two strains, say poky and normal, are crossed, the progeny are like the female (protoperithecial) parent; the male (conidial) parent does not have any influence on the inheritance of these characters, no matter how many times the cross is repeated in successive generations. The maternal inheritance following crosses shows that there are cytoplasmic factors determining these characters, since much more cytoplasm is transmitted through the protoperithecia than through the conidia. However, conidia used as asexual spores do transmit the abnormalities to their progeny, indicating that conidia contain the cytoplasmic factors and transmit them to asexual but not sexual progeny.

Confirmation of this cytoplasmic determination was obtained by study of fused hyphae. In *Neurospora*, heterokaryons—i.e. hyphae containing two types of nuclei, marked with different genes—can be made by hyphal fusions, and cytoplasmic mixing also occurs to some extent. Heterokaryons between poky and normal are initially normal, but eventually show the mutant phenotype. By excision of small fragments of hyphae from such heterokaryons, homokaryons can be re-established, and it is then found that the poky (*mi-1*) phenotype can be obtained in combination with marker genes from either nuclear type (Pittenger, 1956; Gowdridge, 1956), showing that inheritance of poky is not through the nuclei.

A third proof of the cytoplasmic control of poky (and similar variants) in *Neurospora* was obtained by making artificial microinjections of cytoplasmic constituents from one strain into another. This technique was devised by Wilson (1969), who injected purified mitochondria from poky (*mi-1*) into single hyphal segments of normal strains. When the injected segments were excised, cultured and grown up, they were found to resemble the poky type. It seems clear from these experiments that mitochondria from poky strains (as well as from other strains studied by Diacumakos *et al.*, 1965) convert normal hyphae into the variant or defective type.

In general, poky and other similar strains of *Neurospora* are less extreme variants than the petite yeasts. In the poky strains, cytochromes $a + a_3$, though deficient in young cultures, are produced in older cultures, and there is no evidence of gross structural defects in the mitochondria in *Neurospora* like those found in some petite yeasts. All this is understandable considering that *Neurospora* variants as extreme as the petite yeasts would be expected to die.

Drug-resistant mutants in *Saccharomyces cerevisiae* In yeast, many drug-resistant mutants capable of growing in concentrations of certain drugs which inhibit or kill normal, wild-type strains have been obtained. Some of these mutants are controlled by nuclear, others by mitochondrial, genes. Chloramphenicol (CAP) and erythromycin (ERY) are the drugs most often used in this work, but a number of others, such as oligomycin (OLI), spiramycin (SPI), antimycin (ANA), paromomycin (PAR), venturicidin (VEN) and tri-ethyl-tin (TET), have also been studied. These drugs, most of which are used in clinical medicine to kill bacteria, are known to affect certain components of the cell. CAP, ERY, SPI and PAR at appropriate concentrations, inhibit protein synthesis in mitochondrial and bacterial ribosomes, but not in the cytoplasmic ribosomes of eukaryotic cells. On the other hand OLI, VEN and TET inhibit certain components of the ATP-ase complex, while ANA inhibits electron transfer between cytochrome b and cytochrome c.

Linnane *et al.* (1968) found that many drug-sensitive strains cannot grow on a non-fermentable medium (e.g. one with glycerol as a source of energy) containing 0.1 mg/ml ERY. However, by plating cultures of sensitive strains on medium containing up to 8 mg/ml ERY, mutants were obtained, able to grow on concentrations of 8–20 mg/ml ERY.

The inheritance of some of these ERY-resistant mutants is non-Mendelian. Crosses between ERY-resistant and sensitive strains give rise to diploid colonies containing a mixture of resistant and sensitive cells, and following sporulation, ascospore analysis shows either a segregation of 4 resistant:0 sensitive or 4 sensitive:0 resistant spores, depending on whether the diploid cell from which the ascospores developed was resistant or sensitive.

Indications that the non-nuclear genetic factors for ERY-resistance are located in the mitochondria was obtained by inducing (by euflavine treatment) the formation of cytoplasmic petites from ERY-resistant strains (Thomas & Wilkie, 1968b; Linnane *et al.*, 1968). These petites could not be tested directly for ERY-resistance, since they lacked functioning mitochondria. However, the ability of the petites to transmit the genetic factors for ERY-resistance was studied by crossing petites derived from ERY-resistant strains to normal (ρ^+ and ERY-sensitive) strains, and examining the ρ^+ progeny. These were found to be ERY-sensitive, thus making it seem likely that, at the same time that deletions of mitochondrial DNA arise producing petites, the ERY-resistant factors are also lost. Hence the ERY-factors were believed to be situated in mitochondrial DNA.

Linnane *et al.* (1968) found however that not all ρ^- petites failed to transmit ERY-resistance. Whether they did or did not depended on whether the deleted DNA segments contained the locus of the ERY-resistance genes, as shown by more detailed studies on petite production with ethidium bromide (Nagley & Linnane, 1972).

Resistance to other antibiotics (CAP, OLI, PAR, ANA, VEN) was shown by similar methods to be controlled by mitochondrial DNA and confirmation of this has been obtained by later detailed studies whereby the genes concerned have been mapped on the mitochondrial DNA (see p. 30).

In addition to these mitochondrial mutants, resistance to ERY and the other drugs may also arise as a result of mutation in nuclear genes. This will not be considered here.

Drug-resistant mutants in other fungi In *Neurospora crassa*, cytoplasmically inherited drug-resistance mutants have not been described. However, in *Aspergillus nidulans*, mutants resistant to CAP and OLI—and in *Podospora anserina* to CAP—are known and show extranuclear inheritance (Rowlands & Turner, 1975; Belcour & Begel, 1977).

Drug-resistant mutants in *Paramecium* In *Paramecium aurelia* mitochondrial mutants resistant to ERY, CAP and MIK (mikamycin) have been described (Beale *et al.*, 1972; Adoutte & Beisson, 1972; Beale, 1973). These mutants arise spontaneously when paramecia are placed in medium containing one of the drugs at a concentration which inhibits division but does not kill the ciliates. After a period of up to two weeks, during which the animals swim around, getting progressively smaller, a few individuals abruptly resume growth at the normal rate, and are found to be drug-resistant. ERY-resistant mutants arise at a frequency of about 1/1000 cells, or $1/10^6$ mitochondria.

Genetic analysis shows that these drug-resistant mutants are cytoplasmically inherited and proof that the genetic factors concerned are located in mitochondria was supplied by injection of mitochondrial preparations from drug-resistant paramecia into drug-sensitive cells (Knowles, 1974). The injected recipients are placed in drug-containing medium and after an interval of 4–15 days give rise to permanently resistant clones. Under favourable conditions, nearly 100% of the injected cells change. That the process involves a replacement of recipient by donor mitochondria was shown by experiments involving injection of mitochondria from doubly-resistant cells (e.g. to both ERY and MIK) and demonstration of an exact correspondence between the drug-resistance properties of the donors and the transformed recipients, irrespective of genotype or of the drugs in the medium after injection (Beale, 1973).

Further proof that microinjection of drug-resistant mitochondria in these experiments results in a replacement of recipient by donor mitochondria was obtained in some experiments with interspecies transfers, which are successful between certain of the species (syngens) of the *P. aurelia* group. Mitochondrial DNA in different species can be distinguished by restriction endonuclease cleavage studies (Cummings *et al.*, 1976). Following microinjection of species 1 ERY-resistant mitochondria into species 7 cells and growth of the recipients in ERY-containing medium, the mitochondrial DNA of the resistant species 7 cells thus obtained was found to be that of the donors, not the recipients.

Drug resistance in mammalian cells CAP-resistant lines of human (HeLa) cells and mouse tissue culture cells have been obtained (Spolsky & Eisenstadt, 1972; Bunn *et al.*, 1974). In the latter it was shown by a novel technique that the mutation was cytoplasmic. This was done by fusing together enucleated CAP-resistant cells and nucleated CAP-sensitive cells having some chromosomally inherited characteristics (enzyme deficiencies and resistance to other drugs). It was then possible to select out CAP-resistant cells containing chromosomal genes originally associated with CAP-sensitivity. Such recombination in asexually reproducing cells may be taken as proof of cytoplasmic inheritance of CAP-resistance. It was inferred, though not directly proved in this case, that the genetic elements concerned were located in the mitochondria.

Recombination and mapping of mitochondrial genes

Introduction In classical genetics the traditional method of establishing the order of genes along chromosomes is to carry out recombination studies. In higher organisms this is made possible by the regular occurrence of meiosis, while in bacteria and viruses (and to some extent in fungi and some other groups), a number of less regular, 'parasexual' processes have been made use of (Pontecorvo, 1958). Recombination of linked genes in higher organisms occurs as a consequence of crossing-over, a process which involves breakage of homologously paired chromosomes and rejoining of the broken segments in new combinations. This was the classical theory, but later work, particularly that involving study of recombination between very closely adjacent genes, or parts of genes, in microorganisms, showed up the inadequacies of the classical mechanical theory, and other concepts have had to be introduced, such as those involving the action of enzymes on breakage and rejoining of DNA strands, non-reciprocal recombination (conversion), polarity, etc. (see Whitehouse, 1968). The whole problem is still far from clear. However, whenever homologous DNA strands come together there is a likelihood of breakage and rejoining in new combinations.

With mitochondrial genes the chances of detecting recombination might seem to be rather poor, since in the zygotes of many organisms, though not yeast, there is little or no mixing of mitochondria from the two uniting gametes. In higher animals for example, though the sperm may contain vast numbers of mitochondria, few seem to survive for long in the fertilized egg. In rats, Szollosi (1965) found that though sperm mitochondria enter the egg at fertilization, they swell and disintegrate, and rather similar observations have been made with other animals (e.g. a tunicate—*Ascidia nigra* (Ursprung & Schabtach, 1965)).

There is also genetic evidence that sperm mitochondrial DNA is lost after fertilization, at least in interspecific hybrids. The mitochondrial DNA of horses and donkeys has been shown to be different by restriction endonuclease analysis (Hutchison *et al.*, 1974). Mules (offspring of female horses and male donkeys) contain mitochondrial DNA like that of horses, while hinnies (female donkeys × male horses) contain mitochondrial DNA like that of donkeys. It should be added that these hybrids contain nuclear chromosomes and nuclear genes (e.g. for cytochrome c) from both parents. Similarly, in hybrids between the newts *Xenopus laevis* and *X. mulleri*, the mitochondrial DNA is usually like that of the maternal parent, as shown by a technique involving hybridization of mitochondrial DNA from hybrids with cRNA (i.e. RNA produced by transcription of DNA in an *in vitro* system (Dawid & Blackler, 1972)). However, later work showed that recombination between paternal and maternal mitochondrial DNA occurred in some hybrids (Dawid *et al.*, 1974). Finally, in *Paramecium* there is normally little mixing of cytoplasm (including mitochondria) from two partner cells undergoing conjugation, though such mixing may occur under exceptional conditions (Sonneborn, 1950; Beale, 1969).

In yeast, by contrast, formation of the zygotes involves a total mixing of the contents of the two gamete cells, mitochondria from both participate in the growth of hybrid cells and recombinants are common. It has even been suggested (Bolotin *et al.*, 1971) that mitochondria themselves have a kind of sexuality, though this notion, at least in the way originally formulated, is no longer accepted.

In organisms other than yeast relatively little information on recombination of mitochondrial genes is available. In *Neurospora* there are indications from studies on heterokaryons involving slow growing (SG) and normal strains that recombination of the SG factor with other cytoplasmic factors may occur, producing a new type of variant ('slow SG') (Puhalla & Srb, 1967). In *Aspergillus nidulans*, heterokaryons involving three pairs of extranuclear (presumed mitochondrial) genes—for OLI-resistance, CAP-resistance and cold sensitivity—and the corresponding normal alleles were constructed, and recombinants were obtained (Rowlands & Turner, 1975). In *Paramecium aurelia*, on the other hand, no evidence for recombination between the mitochondrial genes for ERY- and MIK-resistance was found (Beale, 1973) or between several other markers (Adoutte, 1977).

Mitochondrial recombination in yeast   The first observations indicating that recombination of mitochondrial genes takes place in yeast were reported by Thomas and Wilkie (1968a), who found that after mating between strains carrying cytoplasmically inherited CAP- and ERY-resistant markers, diploids could be obtained showing the double-resistant (CAP/ERY) type, as well as others sensitive to both drugs. While the former might conceivably be produced by cells containing a mixture of CAP- and ERY-resistant mitochondria, the sensitive cells could only be produced by recombination (or, of course, by mutation, which could be excluded by experiments with haploid control cells).

A noteworthy feature of recombination studies with yeast, reported by many workers, is the occurrence of irregular proportions of the mitochondrial types derived from the two parents used in a mating and of the two reciprocal types of recombinants. These phenomena are denoted polarity of transmission and polarity of recombination respectively (Coen *et al.*, 1970). This is shown especially clearly by the mitochondrial genes for CAP- and ERY-resistance, as can be seen from data in Table 2.2.

In cross (1) there is a large surplus of *cap^s* over *cap^r* genotypes, and in cross (2) the opposite; in cross (1) the recombinant type *cap^{s}ery^s* is much more frequent than the reciprocal recombinant *cap^{r}ery^r*, and in cross (2) the situation is reversed. Other crosses ((3) and (4)), however, involving the same genes (*cap^r/cap^s*, *ery^r/ery^s*), but from different strains, do not show these irregularities, and there is no polarity of transmission or recombination. Moreover, no polarity is shown by certain other genes (e.g. *oli^r/oli^s*, *par^r/par^s*), no matter from which strain they are derived (cross (5)). The recombination value between *oli* and *par* is around 25% and this is close to the maximum for any two genes outside the ω-controlled region (see below) on yeast mitochondrial DNA. Genes separated by this amount of recombination have been described as 'genetically unlinked', though other studies, involving deletions, show that such genes are actually situated on the same circular DNA strand, as discussed later (p. 28).

To account for the polarity phenomenon, Bolotin *et al.* (1971) postulated the existence of a special polarity or sex factor ω (omega), closely linked to the *cap* locus. Yeast strains were classified as ω$^+$ or ω$^-$. Crosses between ω$^+$ and ω$^-$ strains, denoted heterosexual or polar crosses, yield progeny showing the presence of polarity in regard to genes closely linked to ω. Polarity decreases rapidly with distance from ω and the polarity effect has not been found for loci located elsewhere on the genome e.g. *oli* and *par*. Matings between strains containing mitochondria of the same type (ω$^+$ × ω$^+$ or ω$^-$ × ω$^-$), denoted homosexual or non-polar, are also possible and produce recombinants, but usually at a lower frequency, and no polarity is then found.

According to Dujon *et al.* (1974), polarity is due to a non-reciprocal conversion process occurring in hybrids, changing ω$^-$ to ω$^+$ and affecting genes close to the ω locus. Thus, if the allele *cap^r* is on the same DNA strand as ω$^+$, most of the recombinant progeny will be *cap^r*; if *cap^s* is on the same strand as ω$^+$, most will be *cap^s*. Genes at somewhat more distant loci (e.g. *ery*) show the same tendencies but less markedly. Genes which are so far

Table 2.2 Recombination of mitochondrial genes in yeast (after Bolotin *et al.* (1971)—crosses 1–4; and Wolf *et al.* (1973)—cross 5).

Cross	Parents	Progeny (%)			
		$cap^r ery^s$	$cap^s ery^r$	$cap^s ery^s$	$cap^r ery^r$
1 (het)	$\omega^- cap^r ery^s \times \omega^+ cap^s ery^r$	8	54	38	0.13
2 (het)	$\omega^+ cap^r ery^s \times \omega^- cap^s ery^r$	79	3.5	0.4	17.5
3 (homo)	$\omega^- cap^r ery^s \times \omega^- cap^s ery^r$	47.5	41.5	6	5
4 (homo)	$\omega^+ cap^r ery^s \times \omega^+ cap^s ery^r$	44	44.5	7	4.5
		$oli^r par^r$	$oli^r par^s$	$oli^s par^s$	$oli^s par^r$
5 (het)	$\omega^+ oli^r par^s \times \omega^- oli^s par^r$	42.6	31.3	15.4	10.5

Notes (1) ω^+, ω^- polarity or sex factor (see text)
(2) het   heterosexual or polar cross
    homo   homosexual or non-polar cross
(3) cap   locus for chloramphenicol resistance/sensitivity
    ery   locus for erythromycin resistance/sensitivity
    oli   locus for oligomycin resistance/sensitivity
    par   locus for paromomycin resistance/sensitivity

away as to appear 'unlinked', are uninfluenced by ω. However, such genes may nevertheless participate in recombination by a process involving reciprocal exchange or symmetrical conversion.

To account for the upper limit of 25% in recombination values in homosexual crosses and in heterosexual crosses involving genes remote from ω, it is assumed by Dujon *et al.* (1974) that mating occurs repeatedly within a panmictic population of DNA molecules. 'Mating' can therefore occur not merely between two different DNA molecules, derived from genetically diverse parents, but also between two identical DNA molecules and this would of course not lead to any genetically detectible recombination.

Notwithstanding much intensive study, the mechanism of genetical recombination in yeast mitochondrial DNA is still not clearly understood. In some respects, e.g. in the occurrence of multiple rounds of mating, the system is more like that of phage than of bacteria. Whereas in bacteria, conjugation (induced by F or other plasmids) is thought to involve a sequential transfer process, starting at a certain point (the Hfr locus) and proceeding regularly with the elapse of time around the bacterial chromosomes, followed by eventual pairing and recombination of DNA segments, there is no evidence for such a process with mitochondrial DNA. As for the ω factor, nothing comparable seems to have been proposed elsewhere, and in any case the nature and mode of operation of this factor are at present matters for discussion (Perlman & Birky, 1974; Clark-Walker & Gabor Miklos, 1974).

Mapping of yeast mitochondrial DNA A variety of methods can now contribute to the construction of mitochondrial DNA maps in which the positions of individual genes are ordered and related to structural features of the DNA. The main method involves deletions of more or less substantial parts of the DNA. Other methods use either recombination studies or, for certain types of genes, such as those for t-RNAs and r-RNAs, DNA/RNA hybridization methods. Finally, the physical distance separating certain elements on the genetic map may be estimated with reference to the structural map determined by restriction endonucleases. Unfortunately it has not been possible in yeast to isolate the whole of the mitochondrial DNA in one piece, and this has excluded the use of the elegant EM-heteroduplex methods successfully applied to animal mitochondrial DNA (see pp. 29, 30).

In yeast, although recombination studies have been carried out with a number of drug-resistant genes, the difficulty in using these results for the construction of mitochondrial DNA maps is that, due to the occurrence of multiple rounds of mating and hence the opportunity for many recombinational events to take place, genes which are not close together will appear unlinked even though they are located on the same DNA strand. The polarity system also leads to some complications in estimating recombination frequencies in the neighbourhood of ω. On the other hand, complete absence of recombination may be taken as an indication of allelism.

With these limitations, some yeast mitochondrial genes have been assigned to specific groups but these methods give no information of the order of the groups in the genome.

Molloy *et al.* (1971), initiated a different approach. They used spontaneously derived petite stocks in a study designed to give information about the sequence of genes on the mitochondrial genome. The sequence of genes may be ascertained on the principle that two genes close together are more likely to be lost simultaneously when a deletion occurs, than are two widely separated genes. This type of approach has proved useful and several genes have now been placed in order on a circular map of yeast mitochondrial DNA (Schweyen *et al.*, 1976; Trembath *et al.*, 1976). The genes for the r-RNAs and t-RNAs have also been mapped by hybridization between these RNAs and petites containing known marker genes. The larger 23 S r-RNA gene has been located in the region containing the *cap* and *ery* genes (Faye *et al.*, 1974; Linnane *et al.*, 1976) and the smaller 16 S r-RNA gene has been shown to be near the locus of the *par* gene (Faye *et al.*, 1975). Thus the two r-RNA components are produced by widely separated genes. A composite map provisionally embodying data from a variety of sources is shown in Fig. 2.13. A second type of approach has involved the construction of a precise 'physical' map of yeast mitochondrial DNA, based on the specific points at which various restriction endonucleases act (Sanders *et al.*, 1976). The accurate placing of genes on yeast mitochondrial DNA is a very recent area of research and much progress remains to be made. All the techniques used have their disadvantages, in particular the 'scrambling' of mitochondrial genes that occurs during petite formation complicates considerably the use of petites for establishing gene order. However, by using a large number of petites well characterized both genetically and physically it should be possible to arrive at an accurate map of yeast mitochondrial DNA. Work along these lines is already in progress by Nagley *et al.* (see Saccone & Kroon, 1976).

Mapping of mitochondrial DNA in higher animals Although there are at present no data on linkage of mitochondrial genes in animals, some information on the sites of genes for ribosomal and transfer RNAs is available. This comes from studies on the hybridization of the various types of RNA with mitochondrial DNA. Working with HeLa (human) cells, Wu *et al.* (1972) isolated mitochondrial DNA and separated the heavy (H) and light (L) polynucleotide chains. The two types of DNA were then hybridized with three classes of mitochondrial RNA—16S, 12S and 4S— obtained from these cells. The first two are the ribosomal RNAs and the third is thought to comprise mainly transfer RNAs. With the r-RNAs, the positions of homology of the DNA and RNA could be observed directly by the presence of double (heteroduplex) strands, visible by electron microscopy. As for the 4S RNAs, these were first combined with ferritin, an iron-containing protein which shows up clearly in the electron microscope, and then hybridized with the DNA.

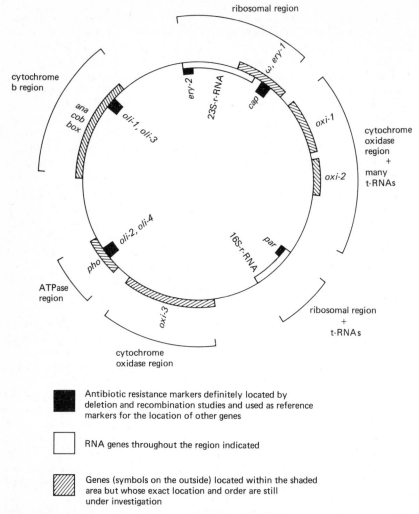

Fig. 2.13 Provisional genetic map of yeast mitochondrial DNA. (After various authors in Saccone and Kroon, 1976.) Gene symbols: *cap*—chloramphenicol resistance; *ery*—erythromycin-resistance; *oli*—oligomycin-resistance; *par*—paromomycin-resistance; *oxi*—cytochrome oxidase; *cob*—cytochrome b; *ana*—antimycin-A; *box*—cytochrome b and cytochrome oxidase; *pho*—oligomycin-sensitive ATPase; ω (omega)—polarity or sex factor.

The H strand bears the two r-RNA genes, which are adjacent and only separated by a small gap, and twelve sites homologous with the 4S-RNAs (one in the gap between the 12S and 16S genes). The L strand bears another seven 4S RNA sites, widely spaced out (Attardi *et al.*, 1976).

In *Xenopus laevis*, Dawid *et al.* (1976), by similar methods, have mapped the positions of the two r-RNA genes in relation to a physical map based on restriction endonuclease studies. Here also, the two r-RNA genes are found to be adjacent and on the H strand. In addition, 15 or 16 sites corresponding to 4S RNAs were found on the S strand, and 5–6 on the L strand.

Conclusion Since the only map of mitochondrial DNA showing the position of genes other than those determining the r- and t-RNAs, is derived from one organism, *Saccharomyces cerevisiae*, it is premature to attempt to draw any firm conclusions as to the significance of the gene order. According to Bernardi *et al.* (1976) there may be as many as 70 gene loci in yeast mitochondrial DNA, and if this is true only a small percentage of the total has been mapped so far. As for the positions of the genes for the two types of r-RNA, it is perhaps surprising to find these genes are widely separated in yeast, but closely adjacent in man and *Xenopus* (and also, in *Drosophila* and *Neurospora* (See Saccone & Kroon, 1976)). In this connection, it is perhaps worth recalling that the two r-RNA genes in bacteria and in eukaryotic nuclei are in adjacent cistrons (Jaskunas *et al.*, 1974; Birnstiel *et al.*, 1971).

The dual genetic control of mitochondria

Introduction Mitochondria develop as a result of the activity of two sets of genes, one in the cell nucleus and the other in the mitochondria. Our aim in this section is to specify the roles of each set of genes, paying particular attention to the mitochondrial genes; and to discuss the interaction between the two genetic systems.

It is relatively easy to identify characters controlled by nuclear genes. Where there are variants, genetic analysis shows them to be inherited according to Mendelian rules. Somewhat less easy is the demonstration of mitochondrial genes: again, where natural variants occur, one can make a genetic analysis, show inheritance through the cytoplasm and in some cases a tie-up with mitochondrial DNA. However, such variants are not easy to obtain, except in unusually suitable organisms such as *Saccharomyces cerevisiae*.

A good deal of information has been derived from experimental procedures other than conventional genetic analysis. Foremost amongst these are methods based on RNA/DNA hybridization. A more indirect approach is to obtain information concerning the sites of synthesis of particular mitochondrial proteins, which may be synthesized either on cytoplasmic ribosomes and then transported into the mitochondria, or on mitochondrial ribosomes. In the former case it is assumed that the genes for those proteins lie in the nuclear chromosomes, and in the latter in the mitochondrial DNA. This has not, however, been directly proved. It is possible, though it may seem unlikely, that nuclear genes control production of messenger RNAs which move into the mitochondria and are translated on mitochondrial ribosomes.

Various inhibitors, such as chloramphenicol, erythromycin, cyclohex-imide, ethidium bromide and others, have been used to obtain information on the sites of synthesis of particular proteins, since some inhibitors (CAP and ERY) inhibit protein synthesis on mitochondrial ribosomes but not on cytoplasmic ribosomes, while others (e.g. cycloheximide) do the reverse. However, as pointed out by Schatz and Mason (1974), these effects may be less specific than is often assumed, and the results of experiments with inhibitors should be regarded as provisional in regard to any conclusions regarding the genetic control of synthesis of particular mitochondrial proteins.

In this section we will start with the mitochondrial DNA itself, and proceed outwards, considering in turn the mitochondrial ribosomes, soluble enzymes and components of the inner membrane. The last stage in this procedure would be to consider which, if any, components of the cell outside the mitochondria altogether are controlled by the mitochondrial genome, but there is little information available about that.

Mitochondrial DNA It is axiomatic in genetics that old DNA determines the nucleotide sequence of new DNA, and this presumably is as valid for mitochondrial as for any other DNA. There is no evidence that the structure of mitochondrial DNA is affected by nuclear DNA. DNA/DNA hybridization studies have failed to reveal the existence of any nucleotide sequences common to nuclear and mitochondrial DNA (Borst, 1972). Nor, so far as we know, does mitochondrial DNA ever become inserted in nuclear chromosomes, as do many bacterial plasmids in bacterial chromosomes.

Crosses between species or sub-species having distinguishable mitochondrial DNAs show that the inheritance of mitochondrial DNA is strictly 'maternal' (see above, p. 25).

Although the specific structure (polynucleotide sequence) of mitochondrial DNA is exclusively governed by that of pre-existing mitochondrial DNA, the functioning of that DNA is probably largely controlled by nuclear genes. This is because such processes as DNA replication, recombination and transcription require the participation of enzymes, which are probably coded by nuclear DNA, though there is little direct proof of this at present. Faye *et al.* (1974) conclude that the basic machinery for mitochondrial DNA replication and its transcription is derived from nuclear genes, since various petite yeast strains containing different deletions of substantial parts of the mitochondrial DNA, nevertheless succeed in replicating and transcribing the remaining DNA.

In *Tetrahymena* (and presumably in other organisms) there are two different DNA polymerases, one for nuclear DNA and the other for mitochondrial DNA. Synthesis of mitochondrial DNA polymerase was shown to be actually enhanced by treatment with certain substances, such as ethidium bromide which blocks mitochondrial, but not nuclear, DNA activity. Hence it is inferred that the mitochondrial DNA polymerase is coded by nuclear DNA (Westergaard *et al.*, 1970).

Thus, with the extremely limited information at present available, one can draw the tentative conclusion that the functioning of mitochondrial DNA, like that of bacterial plasmids (see p. 70), is mainly controlled by genes not within these extranuclear DNAs.

Mitochondrial ribosomes As previously stated (p. 14) mitochondria contain ribosomes consisting of two sub-units, each containing ribosomal RNA and a large number of ribosomal proteins. All the evidence indicates that the r-RNAs are coded by mitochondrial DNA. Unlike bacteria, which are thought to contain about six sets of r-RNA genes (Jaskunas *et al.*, 1974), and eukaryotic nuclei, which contain many hundreds of such sets (Birnstiel *et al.*, 1971), each molecule of mitochondrial DNA has just one pair of r-RNA genes. Where these are close together (*Xenopus*, HeLa), there is also a site coding for a 4S RNA in the spacer between them, but it is not known if this is analogous to the 5S r-RNA gene in bacteria, chloroplasts or eukaryotic nuclei.

As regards the fifty or more mitochondrial ribosomal proteins, it is frequently stated that they are coded by nuclear genes, synthesized on cytoplasmic ribosomes and then transported into the mitochondria to unite with the mitochondrial r-RNA. This view is based on the finding that synthesis of these proteins is not inhibited by CAP, but is inhibited by cycloheximide, in *Neurospora*, yeast and HeLa cells (Schatz & Mason, 1974). Moreover in *Neurospora*, Lizardi and Luck (1972) separated 53 proteins, and showed that all were inhibited by anisomycin (an inhibitor of cytoplasmic ribosomes), but none by CAP. Finally, none of these proteins were found to incorporate labelled amino acids in isolated mitochondria, i.e. in the absence of cytoplasmic ribosomes. It should however be mentioned that some discrepancies have been reported in parallel experiments with inhibitors in yeast, *Tetrahymena*, and *Neurospora* (Schatz & Mason, 1974). Moreover, in a recent publication on *Neurospora* (Lambowitz *et al.*, 1976), it was reported that one ribosomal protein (denoted S-4a) was synthesized within the mitochondria, and this protein was deficient in the poky mutants.

Unfortunately direct genetic analysis, like that done with variants of ribosomal proteins in bacteria (Jaskunas *et al.*, 1974), has not been carried out with mitochondrial ribosomal proteins.

In *Paramecium aurelia* Tait (1972) showed that an ERY-resistant mutant had altered mitochondrial ribosomes, and tentative evidence was obtained showing that a mitochondrial ribosomal protein in the mutant had altered electrophoretic properties (Beale *et al.*, 1972). Moreover, Tait *et al.* (1976a) have shown that some mitochondrial ribosomal protein differences between two species (syngens) of the *P. aurelia* group, as detected by immunological methods, are coded by mitochondrial DNA.

In *Xenopus*, Leister and Dawid (1975) studied differences between mitochondrial ribosomal proteins in the two species *X. laevis* and *X. mulleri*, and found that inheritance of four variant types was maternal, thus implying, though not conclusively proving, mitochondrial coding of some of these proteins.

To sum up, the question of the coding for mitochondrial ribosomal proteins cannot be considered as decisively settled. The studies with inhibitors have given contradictory results: in some organisms most if not all of these proteins are the products of cytoplasmic protein synthesis, whereas in other organisms a proportion of the proteins are synthesized on mitochondrial ribosomes. In any case, as stated previously, the genetic significance of these experiments using inhibitors should be interpreted with caution. Finally, it may well be that the situation is not the same in different organisms.

Mitochondrial transfer RNAs In regard to mitochondrial t-RNAs, the evidence is predominantly in favour of coding by the mitochondrial DNA. Data from various organisms are summarized in Table 2.3. *Tetrahymena pyriformis* appears to be exceptional in showing indication of nuclear control of some mitochondrial t-RNAs (Chiu *et al.*, 1975).

Table 2.3 Present estimate of number of mitochondrially coded t-RNAs (provisional).

Organism	No. of t-RNAs
Saccharomyces cerevisiae[1]	16
Saccharomyces carlsbergensis[2]	25
Neurospora crassa[3]	40
Xenopus laevis[3]	22
Human (*HeLa*)[4]	19

References: *1* Rabinowitz *et al.* (1976); Bolotin-Fukuhara *et al.* (1976)
        2 Reijnders & Borst (1972)
        3 Blossey & Küntzel (1972)
        4 Attardi *et al.* (1976)
        5 Dawid *et al.* (1976)

Mitochondrial RNA polymerase As regards mitochondrial RNA polymerase, this was considered by Barath and Küntzel (1972) working with *Neurospora*, to be translated not on mitochondrial but on cytoplasmic ribosomes, since ethidium bromide, which inhibits the mitochondrial system, produced an increase in amount of this enzyme. Moreover, various petite yeasts, though incapable of carrying out mitochondrial protein synthesis, nevertheless contain mitochondrial RNA polymerase (Wintersberger, 1970). It is therefore assumed that the transcription of mitochondrial DNA is controlled by enzymes coded by nuclear, not mitochondrial, DNA. There is, however, no direct genetic evidence for this.

Soluble enzymes It is generally believed that the soluble enzymes in the mitochondrial matrix are all determined by nuclear genes and translated on cytoplasmic ribosomes, though genetic data are available for only a few out of the vast numbers of such enzymes. Variants of mitochondrial malate dehydrogenase (MDH) show Mendelian inheritance in various organisms, such as *Neurospora crassa* (Munkres *et al.*, 1965); mouse (Shows *et al.*, 1970); maize (Longo & Scandalios, 1969), and man (Davidson & Cortner, 1967). In *Paramecium aurelia*, Tait (1968, 1970) has shown nuclear gene control of the mitochondrial enzymes isocitrate dehydrogenase and β-hydroxybutyrate dehydrogenase. Genetic data are available for some seven or eight other mitochondrial enzymes in various organisms, and all have been found to be controlled by the nuclear, not the mitochondrial, genome.

Even where strictly Mendelian methods cannot be used, due to the impossibility of hybridizing variant forms of non-interbreeding species, there may be evidence for nuclear control. Man-mouse somatic hybrids have been used in this way (van Heyningen *et al.*, 1973, 1975). In these cell hybrids the human chromosomes are gradually lost, though individual chromosomes may remain; human mitochondrial DNA is lost completely. In some hybrid cells the human forms of the mitochondrial enzymes citrate synthase and malate dehydrogenase were present, and the latter was correlated with retention of the human chromosome No. 7. Hence the gene for this enzyme was located on that chromosome. By similar methods the genes for mitochondrial isocitrate dehydrogenase and fumarate hydratase have been shown to be located on nuclear chromosomes (Bruns *et al.*, 1976; Craig *et al.*, 1976).

In *Paramecium* the technique of micro-injection of mitochondria from one species (or syngen) to another has been used. The mitochondrial enzyme fumarase has different electrophoretic forms in species 1 and 7 of *P. aurelia*, and it was shown that the type of enzyme produced was correlated with the nucleus, not the mitochondria, of cells into which mitochondria has been injected (Knowles & Tait, 1972).

Cytochrome c The chemical structure of cytochrome c is wholly controlled by the nuclear genome. Many mutants containing variants of cytochrome c have been studied, and inheritance of them all is Mendelian. The amino acid sequence has been worked out, and shown to be controlled by genes at a specific chromosomal locus in yeast (Sherman & Steward, 1971). Furthermore petite yeasts, which are incapable of carrying out protein synthesis in mitochondria, and even lack mitochondrial ribosomes, nevertheless contain nearly normal amounts of cytochrome c; and finally normal yeasts continue to synthesize cytochrome c in presence of inhibitors of mitochondrial protein synthesis.

Inner membrane proteins The mitochondrial inner membrane is the main site of important respiratory processes which involve an exceedingly complex series of biochemical reactions. A few components participating in these reactions have been subjected to genetic study, and have been shown, at least in part, to be under the control of the mitochondrial genome. These are (1) the oligomycin-sensitive ATP-ase complex; (2) the cytochrome oxidase complex, and (3) cytochrome b (Tzagaloff, 1975; Schatz & Mason, 1974).

The oligomycin-sensitive ATP-ase complex consists of three parts: (a) F_1 ATP-ase, which is part of the catalytic unit, and comprises five polypeptides; (b) OSCP (=oligomycin-sensitivity-conferring protein), which links F_1 to the membrane fraction, and comprises one polypeptide, and (c) the inner membrane fraction, which has no known enzymatic function and comprises four polypeptides. The arrangement of the three components is shown diagrammatically in Fig. 2.14.

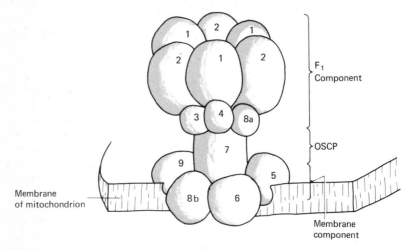

Fig. 2.14 The mitochondrial ATPase. This shows the probable location of the nine subunits in the ATPase complex. Subunits 1, 2, 3, 4, 7 and 8a are synthesized on cytoplasmic ribosomes and are probably coded by nuclear DNA. Subunits 5, 6, 8b and 9 are synthesized on mitochondrial ribosomes and are probably coded by the mitochondrial DNA. These mitochondrial subunits are closely associated with the inner mitochondrial membranes. (This diagram is taken with thanks from a sketch by Dr J. Cosson.)

From studies in yeast with the inhibitors cycloheximide and CAP, it is believed that F_1 and OSCP are synthesized entirely on cytoplasmic ribosomes and imported into the mitochondria, while part of the membrane fraction is synthesized on mitochondrial ribosomes. Furthermore reconstruction experiments with components from wild type and OLI-resistant mutants have shown that OLI-resistance is due to a modification of the membrane fraction, not of F_1 or OSCP (notwithstanding the name 'OSCP'). Since these OLI-resistant mutants are controlled by mitochondrial genes, we have genetic proof that the membrane fraction is at least partly controlled by mitochondrial genes (Griffiths & Houghton, 1974).

The cytochrome oxidase complex comprises seven polypeptides, of which the three larger ones have been shown, again by inhibitor studies, to be synthesized on mitochondrial ribosomes, both in yeast and *Neurospora* (Schatz & Mason, 1974), and the three smaller polypeptides are synthesized on cytoplasmic ribosomes. However, some nuclear (*pet*) mutants cause the loss of one or more of the mitochondrially-made polypeptides (Schatz, 1975).

Cytochrome b comprises a haem group and two polypeptides of nearly equal size. Inhibition studies in *Neurospora*, with CAP and cycloheximide, have shown that at least one of these polypeptides is synthesized on mitochondrial ribosomes.

As stated earlier, mitochondrial variants of *Neurospora* (poky, *mi*, etc.) and yeast (ρ^- petites) have long been known to be deficient in various components of the respiratory system, such as cytochrome $a + a_3$, b, cytochrome oxidase etc. However, since many of these variants have suffered substantial losses of mitochondrial DNA, they are unsuitable for demonstrating a direct connection between particular mitochondrial genes and particular components of the inner membrane. Indeed, since in some cases the entire mitochondrial protein synthesizing system, including the ribosomes, is defective or lacking, the effect of these losses of DNA on the membrane proteins could be quite indirect. For example, Rifkin and Luck (1971) showed that in poky (*mi-1*) *Neurospora*, there is a defect in the synthesis or assembly of the mitochondrial small (19S) r-RNA sub-unit, and a close correlation between deficiencies of the sub-unit and of cytochromes a and b, at least under certain conditions. As indicated above, the primary effect of the poky gene or genes is probably on the ribosomal protein(s), and not on the cytochromes.

Recently many individual mutations in yeast, denoted *mit*⁻, have been obtained, and these may involve small regions of the mitochondrial DNA, with specific effects on membrane proteins. Some of these mutants (e.g. *oxi-1*, *cob*, etc.) are indicated on the yeast mitochondrial genetic map shown in Fig. 2.13. It is clear that various mutations which can be mapped at specific sites on the mitochondrial DNA lead to inactivation of particular inner membrane components, such as cytochrome oxidase, ATP-ase, cytochrome b, and others. Some mutants (*box*) confer simultaneously a deficiency in both cytochrome b and cytochrome oxidase but the exact nature of such mutations is controversial (Pajot *et al.*, 1976; Tzagoloff *et al.*, 1976). Intensive study of this matter is at present in progress.

In *Paramecium aurelia* it has been shown by Tait *et al.* (1976b) that a hydrophobic protein fraction derived from mitochondrial membranes varies in different species (syngens), as shown immunologically, and that this variation is in part controlled by the mitochondrial genome. With this material it is possible that future work may reveal the existence of variations in specific proteins, making possible the identification of 'structural' genes for mitochondrial proteins.

Some further mitochondrial/nuclear interactions in Paramecium

In this section we mention two further illustrations of interactions between the mitochondrial and nuclear genetic systems in *P. aurelia*. The first example concerns a slow growing mutant cl_1, deficient in cytochrome oxidase. This mutant differs from wild type due to a nuclear mutation cl_1 and also a heritable modification of the mitochondria. The nuclear mutation cl_1 causes severe disorganization of wild type mitochondria (Sainsard, 1975, 1976). A spontaneous revertant occurred (denoted cl_1-*su*) in which the generation time was found to be nearly wild type, though the cytochrome oxidase defect and thermosensitivity characteristic of the original mutant remained. Genetic analysis showed that the reversion was due to a change in the mitochondrial genome, and is therefore described as a 'suppressor'. It is presumed to act through a functional interaction between two modified products, the one under nuclear, the other under mitochondrial, control.

The second example concerns the compatibility of mitochondria taken from one species (syngen) of the *P. aurelia* group and transferred by microinjection to another (Beale & Knowles, 1976). Where donor and recipient cells are closely related, such transfers sometimes succeed. For example, ERY-resistant mitochondria from species 1 may be readily transferred to species 7 and displace the original species 7 mitochondria, though the reciprocal transfer of species 7 mitochondria to species 1 has not been successfully done. The mitochondrial DNA in the 'hybrid' (sp. 1 mitochondria + sp. 7 nucleus) was shown by studies with restriction endonucleases to be like species 1 mitochondrial DNA and the gene for ERY resistance was also transferred. Nevertheless these 'hybrid' mitochondria were incapable of being transferred back to species 1, and it is concluded that the nuclear genome of species 7 affects some component (possibly in the mitochondrial membrane) making the mitochondria incompatible with species 1 cells. These interactions between nuclear and mitochondrial factors show interesting analogies with certain other incompatibility systems, such as those involved in cytoplasmic male sterility in plants (see p. 110).

Conclusion The present situation regarding the genetic control of different mitochondrial constituents is summarized in Table 2.4. This may create the impression—which is very unlikely to be true—that the nuclear and mitochondrial genomes make approximately equal contributions to the synthesis of mitochondrial constituents. Probably 90% of mitochondrial proteins are synthesized in the cytoplasm. Hence, while both genomes are vitally necessary, the number of nuclear genes controlling mitochondrial characteristics is likely to be vastly greater than the number of mitochondrial genes.

Kinetoplasts

The mitochondria of certain flagellate protozoa are peculiar in their association with a conspicuous DNA-containing structure known as the kinetoplast. Although its genetic role is unknown, the kinetoplast may be considered as an example of an extranuclear organelle showing autonomous reproduction, and we therefore give a short account of it here. (For further details see Vickerman, 1965; Simpson, 1972; Fouts *et al.*, 1975; Saccone & Kroon, 1976). This group of protozoa (the Kinetoplastidae) include such important disease-causing organisms as *Trypanosoma* and *Leishmania*, as well as some free living forms, e.g. *Bodo*.

The kinetoplast is defined by Simpson (1972) as 'that portion of the mitochondrion of Kinetoplastidae containing the fibrous mass of mito-chondrial DNA'. The kinetoplasts are usually situated near to the base of the flagella, but are distinct from the flagellar basal bodies or kinetosomes. Each protozoan cell contains a single kinetoplast, and probably a single mitochondrion (Vickerman, 1965). Electron microscope studies show that the mass of kinetoplast DNA lies in the mitochondrial matrix, close to the inner membrane. In these protozoa the mitochondria are peculiar, not only in their very large content of DNA, but also in the development of long tubular extensions throughout the cell, some of which may be quite remote from the kinetoplast.

Kinetoplasts are easily seen in stained preparations by light microscopy. They are so large as to contain an appreciable proportion of the DNA in the cell—e.g. in *Leishmania tarentolae* 17–20%, and *Crithidia luciliae* 20–30%—while in *Bodo* the kinetoplast may be larger than the nucleus. However kinetoplast DNA can be distinguished from that of the nucleus by its sedimentation in caesium chloride gradients. For example in *Crithidia luciliae* the buoyant density of kinetoplast DNA is 1.705g/cm^3, and of nuclear DNA it is 1.717g/cm^3 (Fouts *et l.*, 1975).

Electron microscope studies show the structure of kinetoplast DNA to very unusual (Fig. 2.15). It consists of a vast number (about 10^4 per cell) of interlocked 'minicircles', the diameter of which varies according to the species of protozoan. It is $0.25 \,\mu\text{m}$ in *Trypanosoma brucei* and $0.8 \,\mu\text{m}$ in *Crithidia luciliae*. In addition, a proportion of the kinetoplast DNA is in the form of larger (or 'maxi') circles, approximately $11\mu\text{m}$ in length (Steinert & van Assel, 1975; Kleisen *et al.*, 1976) and these are comparable in size with the mitochondrial DNA in other organisms (see p. 13).

In view of the extremely small size of the minicircles, which are roughly comparable with some small bacterial plasmids, very little genetic information can be present. Assuming a molecular weight of 0.56×10^6 daltons (equivalent to 838 nucleotides), at most one small protein of molecular weight 20 000 or a small r-RNA molecule could be produced. The 'maxicircles' could of course contain as many genes as the mitochondrial DNA in some other organisms. No information about transcription products of kinetoplast DNA has been reported.

Some indication of the function of kinetoplast DNA can be obtained from the study of cells lacking kinetoplasts (denoted a- or dys-kinetoplastic forms). These occur spontaneously or can be obtained following treatment with various dyes, such as acriflavine and ethidium bromide. Loss of

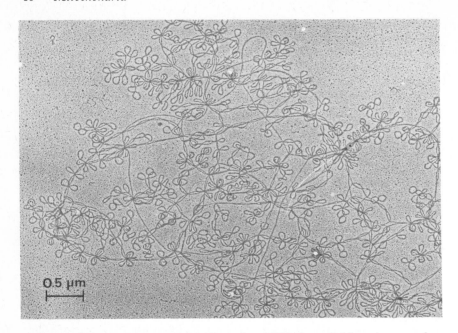

Fig. 2.15 Electron micrograph of kinetoplast DNA from *Leishmania tarentolae* spread by the formamide technique. The minicircles, with some longer components are organized into large associations, each containing about 10 000 minicircles. The whole complex has a molecular weight of around 10^{10} daltons. (With permission from Price, DiMaio and England (1976), in *The Genetic Function of Mitochondrial DNA*, eds. C. Saccone and A. Kroon, and North Holland, Amsterdam.)

kinetoplasts has lethal consequences for the blood forms of some species of flagellate but not for others, e.g. *Trypanosoma brucei*, which remains normal in respect of motility, pathogenicity and general morphology. Only those forms which can respire in mammalian blood by a non-mitochondrial (1-α-glycerophosphate oxidase) system can survive loss of kinetoplast DNA, and the dyskinetoplastic condition is lethal in the insect stages of the life cycle in all species. There is also some evidence that dyskinetoplastic forms have mitochondria which are swollen and lack cristae and may be defective in respiratory enzymes and cytochromes; but as pointed out by Simpson (1972), it is not certain whether the relation between loss of kinetoplast DNA and the defects in mitochondria is one of cause and effect.

It is concluded that kinetoplast DNA provides some necessary components for mitochondrial development in the stage in the insect vector and during the vertebrate stage in those species of *Trypanosoma* and *Leishmania* which rely on a mitochondrial type of respiration, but that kinetoplast DNA is dispensable, and possibly non-functioning, in the mammalian bloodstream phase of *T. brucei* and similar forms. Dyskinetoplastic forms may thus be regarded as analogous to the ρ^0 petite strains of yeast. Some species, e.g. *T. equinum*, which lack an insect stage altogether are 100% dyskinetoplastic.

By comparison with mitochondria in other groups of organisms, those of the Kinetoplastida are very distinct, above all in the enormous amount of the kinetoplast DNA and in the presence of the minicircles, whose function is at present quite unknown. Unfortunately nothing is known of the genetics of these organisms.

Conclusion

It is probably unwise to attempt to draw general conclusions about the significance of the mitochondrial genetic system at present, since so much of our knowledge about it is derived from only one species, namely *Saccharomyces cerevisiae*, and this species is unusual amongst living organisms in being able to survive even when the mitochondria are non-functional.

However, some characteristics are common to mitochondria in a sufficient variety of organisms to make it worth while attempting some generalizations. The first is that all functioning mitochondria contain DNA, although the unit amount is very small, equivalent to that in some plasmids or viruses, and much less than that in bacteria or endosymbionts. Since it is always present, however, this small amount of DNA is presumably essential. Normal mitochondria in all organisms contain a functional protein-synthesizing system, and certain parts of this system—primarily the r- and t-RNAs—are coded by the mitochondrial DNA. Presence of this protein-synthesizing system implies that some proteins are necessarily synthesized on it. At this point one can only speculate (Table 2.4). It has been suggested that some of the respiratory system essentially involves the highly hydrophobic environment of the inner mitochondrial membranes, though the precise role of such a hydrophobic environment has not been established. Hydrophobic proteins may be relatively immobile in the aqueous medium of the cytoplasm or mitochondrial matrix and difficult or impossible to transport through membranes. Hence they may require to be synthesized close to the site where they will finally rest. It is possible that this is achieved by having some of the mitochondrial ribosomes attached to the inner surface of the mitochondrial inner membrane. No reasons, however, seem to have been advanced for the need to have the mitochondrial r- and t-RNAs transcribed from mitochondrial DNA.

Yet a Darwinian biologist would naturally try to find some advantage in so seemingly awkward a mechanism. The two genetic systems—nuclear and mitochondrial—interact intimately at several levels: firstly, the level of the mitochondrial DNA, whose structure is controlled by itself, but whose activity is controlled by products of the nuclear system; secondly, the level of the mitochondrial ribosomes, part of whose structure is controlled by each system and whose activity also by both; and thirdly, the level of the inner membranes, which contain an integrated patchwork of products of the

Table 2.4 Genetic control of mitochondrial constituents (provisional summary)

Constituent	Controlled by nuclear genome	Controlled by mitochondrial genome
1 Mitochondrial DNA-polynucleotide structure	–	✓
2 Factors controlling replication, recombination, transcription of mitochondrial DNA	✓	?
3 Protein synthesizing system in mitochondria:		
r-RNA	–	✓
ribosomal proteins	✓	?
t-RNA	✓	✓
4 Soluble enzymes	✓	–
5 Other proteins:		
ATP-ase	✓	✓
Cytochrome oxidase	✓	✓
Cytochrome b	✓	✓
Other cytochromes	✓	?

nuclear and mitochondrial genetic systems. All this must require an extraordinary degree of mutual adaptation and correlated evolution.

Our knowledge of the actual machinery of the mitochondrial genetic system is still extremely slight. In yeast, recombination seems to be so frequent an event that practically all mitochondrial genes are randomly assorted. Study of mitochondrial recombination in other organisms is greatly needed. In *Paramecium* there is probably no recombination at all. We therefore have the two extremes of universal, frequent recombination in one organism and none at all in another. If that is the situation, it is very difficult to appraise its significance.

We will discuss the various speculations concerning the evolution of mitochondria, together with that of other organelles, in the final chapter.

3 Chloroplasts

Introduction

Light-driven chemical reactions have played a very important role since the origin of life itself. In particular, photosynthesis, the production of carbohydrates from CO_2 and H_2O is essential for the continued existence of life as we know it. There is a wide range of organisms which are able to convert solar radiation into chemical energy. These vary in size from the giant redwood trees of California to photosynthetic bacteria of only 100 millionth of this size.

The genetic interactions between different components of photosynthetic systems are extremely varied. In some systems, like the lichen species *Collema*, the blue-green algae *Nostoc* is only loosely associated with the hyphae of its mycobiont. In this case the photosynthetic units are distinctly separated from the fungus which forms the other part of the lichen (Ahmadjian & Hale, 1973).

In plants the association is much closer. The photosynthetic unit is completely enclosed in the plant cell and dependent on the plant cell for the manufacture of many components. The chloroplast is nevertheless separated from the rest of the cell by a double semi-permeable membrane.

Chloroplasts contain all the enzymatic machinery necessary for the complex process of photosynthesis. The genetic information determining the structure of these components is partitioned between the nuclear genome of the 'host' cell and the genome of the chloroplast itself.

It is the object of this chapter to consider in detail the nature of the interactions between chloroplast and host cell and in particular the function of the chloroplast genome. The chapter is divided into three main sections. The first section is a brief description of the chloroplast in terms of its structure and macromolecular components. The second section contains a general description of the inheritance of plastid variation in higher plants and *Chlamydomonas*. The third section attempts to identify the relative contributions of the nuclear and chloroplast genomes to chloroplast biogenesis.

Structure and macromolecular components

Functional chloroplasts are found in the leaves and shoots of the overwhelming majority of higher plants, but unlike mitochondria, which are found in almost all cells, functional chloroplasts exist only in photosynthetic tissues. The actual number of chloroplasts per cell is extremely variable, ranging from several hundred in some larger algal cells to one in some *Chlamydomonas* species. Higher plant mesophyll cells contain around 30.

Membrane structure Chloroplasts in both algae and higher plants have similar gross structure (Figs. 3.1, 3.2, 3.3 and 3.4). The functional units of

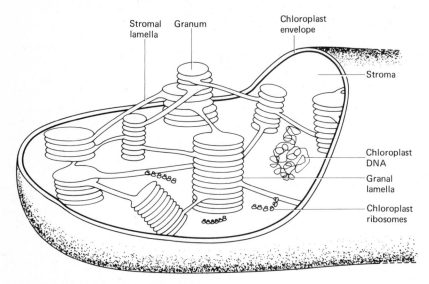

Fig. 3.1 Diagram of part of a chloroplast. Removal of part of the chloroplast envelope shows the main structural features of a mature chloroplast; stroma, grana and stromal lamellae. The DNA is probably attached to the membranes, but ribosomes are often found free in the stroma.

photosynthesis are arranged in closed flattened vesicles called thylakoid disks (Wehrmeyer, 1964). These disks have clearly defined inner and outer surfaces and contain the intralamellar space. The actual arrangement of the thylakoids varies considerably; they may exist either singly, in parallel arrays or stacked one over the other to form grana. In mature chloroplasts the disks are often joined by a complex system of tube-like structures called stromal lamellae. The thylakoid disks and stromal lamellae, surrounded by the stroma or matrix, are contained within the double membrane of the chloroplast envelope. Observations of chloroplasts *in vivo* by micro-ciné photography, have shown that the boundary membrane is very motile and apparently fuses with other chloroplasts. This is sometimes called the motile phase. The lamellar structures contained within are more static (Wildman *et al.*, 1962). A second cinematographic study recorded the division of several chloroplasts in the alga *Nitella*, showing that binary fission is probably the normal mode of chloroplast replication (Green, 1964). There is now

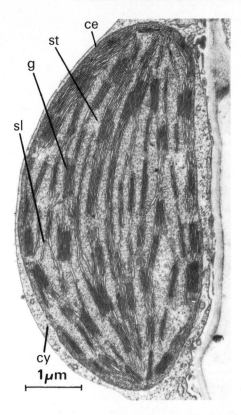

Fig. 3.2 Electron micrograph of mesophyll chloroplast of *Zea mays*. It shows the complex membrane structure formed from grana (g) and stromal lamellae (sl) surrounded by the lighter coloured stroma (st). The chloroplast is separated from the cytoplasm (cy) by the chloroplast envelope (ce). (By kind permission of J. Braugeon.)

considerable cytological evidence that chloroplasts do divide in a wide range of plants (Kirk & Tilney-Bassett, 1967).

Almost all the components characteristic of photosynthetic electron transport are located in the lamellar membranes. The matrix or stroma surrounding these membranes contains the 'soluble' enzymes of CO_2 fixation and the Calvin cycle. Chloroplast DNA, RNA and the chloroplast ribosomal proteins are also found in the matrix although the DNA and chloroplast ribosomes may sometimes be closely associated with the membrane systems (Kirk & Tilney-Bassett, 1967).

Ribosomes and RNAs A large proportion (between 20 and 50%) of the ribosome population of photosynthetic cells consists of chloroplast ribosomes assembled and located within the chloroplast (Boardman, 1967). There is now general agreement that most plant chloroplast ribosomes have

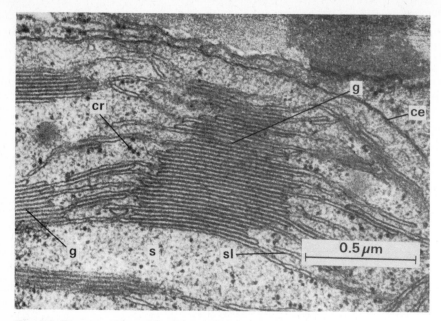

Fig. 3.3 Electron micrograph of part of a chloroplast from an anther of *Zea mays*. It shows a large well developed granum (g) connected to other grana by stromal lamellae (sl). Chloroplast ribosomes (cr) are seen in the stroma (s). (ce—chloroplast envelope). (By kind permission of B. Stevens.)

Fig. 3.4 Electron micrograph of a section through a cell of *Euglena gracilis*. It is possible to see a chloroplast (c) and mitochondria (m). Part of the nucleus takes up the lower section of the figure. In this algae 'grana' (g) are much simpler than in higher plants but often run the length of the chloroplast. (By kind permission of G. Ledoigt.)

a sedimentation value of 67–70S (Ellis & Hartley, 1974), see Table 3.1. Chloroplast ribosomes are thus more uniform in S value than are mitochondrial ribosomes. Chloroplast ribosomes are slightly smaller than cytoplasmic ribosomes, and studies on the subunit sizes have shown that whereas the two chloroplast subunits sediment at between 33–50S and 28–35S the cytoplasmic ribosomal subunits sediment at between 55–67 and 30–40S.

Table 3.1 Sedimentation values of chloroplasts and cytoplasmic ribosomes and their subunits (taken in part from Ellis & Hartley, 1974).

Species	S value of chloroplast ribosome	S values of chloroplast ribosomal subunits	S value of cytoplasmic ribosome	S values of cytoplasmic ribosomal subunits
Algae:				
Euglena gracilis	70	50, 30	88	67, 46
Chlamydomonas reinhardi	68	33, 28	80	—
Higher Plants:				
Nicotiana tabacum	70	50, 35	80	55, 30
Spinacea oleracea	70	50, 30	80	—
Phaseolus vulgaris	70	—	80	—
Pisum sativum	70	45, 32	80	56, 40

N.B. Different workers obtain slightly different values for the same organism; thus these figures are only approximate.

Chloroplast ribosomes are sensitive to antibacterial antibiotics such as chloramphenicol, erythromycin, lincomycin, neomycin, spectinomycin and streptomycin, as shown by studies of poly-U directed phenylalanine incorporation and by radioactive antibiotic binding. It seems that both nuclear and extranuclear genes can determine the resistance of chloroplast ribosomes to these antibiotics. This question will be discussed in more detail later (p. 60).

The sedimentation values for chloroplast ribosomal RNAs isolated from a wide range of plants are close to 23S and 16S. These differ from the values of cytoplasmic ribosomal RNAs which vary from 25S and 20S in *Euglena* to 25S and 18S in higher plants. Several workers report the presence of 5S RNA in chloroplast ribosomes. This is in contrast to the apparent absence of such an RNA in mitochondria. The chloroplast ribosomal proteins of *Chlamydomonas* have been compared to those of the cytoplasmic ribosomes by two-dimensional electrophoresis (Hanson *et al.*, 1974). The chloroplast ribosome contains about 26 proteins in the large subunit and 22 proteins in the small, in contrast to the numbers of cytoplasmic ribosomal proteins, which are 39 and 26, respectively. It cannot be ruled out at present that the chloroplast and cytoplasmic ribosomes have a few proteins in common.

Chloroplasts have been shown to contain specific chloroplast transfer RNAs not found in the cytoplasm. In addition, chloroplasts contain mRNA which is translated into proteins on the chloroplast ribosomes (Sager, 1972; Hartley *et al.*, 1975).

Chloroplast DNA and its replication Chloroplast DNA in higher plants has a buoyant density of between 1.695–1.697 and a GC content of between 36 and 41% (Kirk, 1971). It renatures much more rapidly than the more complex nuclear DNA and has a kinetic complexity of around 10^8 daltons (Ellis and Hartley, 1974). There is increasing evidence that chloroplast DNA isolated from both higher plants and algae is circular with a length of about 40 μm (Manning *et al.*, 1972). This agrees quite well with the kinetic complexity previously estimated by renaturation.

Chloroplasts like mitochondria seem to be polyploid, containing 10–60 copies of a circular genome. These DNA molecules are arranged typically in 5–6 regions within the chloroplast (Herrmann, 1973). Chloroplast DNA from pea and maize plants has been found to contain both Cairns type and rolling circle replication intermediates (Kolodner & Tewari, 1975).

Chloroplast proteins Chloroplasts contain a great number of different types of proteins. Many of these, like the enzymes of the Calvin cycle, are present in the stroma or matrix in a soluble or semi-soluble form. Others, such as the components of photosystems I and II, are in hydrophobic complexes in the membrane system. At present our knowledge of the chloroplast proteins is less complete than of the mitochondrial proteins and naturally this makes precise genetic studies on the location of genes coding for chloroplast proteins difficult.

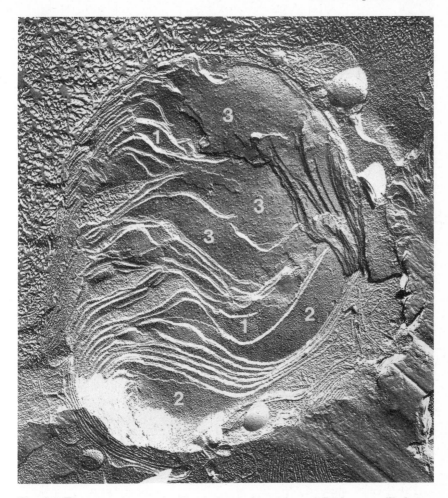

Fig. 3.5 Electron micrograph of an isolated chloroplast of *Euglena gracilis* freeze fractured in Tris-Mg^{++} medium. In this treatment fractures occur between the two halves of membranes, revealing hydrophobic lipoprotein complexes as particles embedded in the membrane. Three fracture faces are recognizable here and can be distinguished by the size and concentration of particles found on their surfaces. ($\times$ 24 000) (By kind permission of M. Lefort-Tran.)

In contrast to mitochondria, chloroplasts contain a considerable amount of one particular enzyme, 'Fraction I' protein. This is the enzyme ribulose-1-5-diphosphate carboxylase, which is responsible for the fixation of carbon dioxide in photosynthetic organisms. This is a key enzyme in photosynthesis and will be discussed in more detail later.

In summary, chloroplasts are complex membrane structures containing all the components required for protein synthesis. They possess their own DNA and RNA and protein-synthesizing system, all of which are distinct from those found in the rest of the cell.

Genetics of chloroplasts

Introduction The earliest reports of non-Mendelian inheritance are found in two studies of variegation in higher plants first published in 1909 by Baur and by Correns (see Kirk & Tilney-Bassett, 1967). Some of the green and white variegated patterns found in plants were shown to be caused by factors inherited in a non-Mendelian manner. Since that time much work has been done on the analysis of variegation in higher plants. This has revealed something of the complex nature of the transmission of chloroplast genetic factors from one generation to the next. However, the difficulty of obtaining specific plastid mutations has limited the study of genetic interactions within the chloroplast genome of higher plants. For this reason our knowledge of the behaviour of chloroplast genes comes mainly from studies using the single celled alga *Chlamydomonas*. This section contains a brief description of the genetic interactions of the chloroplast and host genomes of higher plants, followed by a discussion of the extensive genetic studies of the *Chlamydomonas* chloroplast.

Higher Plants Both the early reports, that of Correns on *Mirabilis*, and that of Baur on *Pelargonium*, were concerned with the inheritance of leaf variegation, caused by a patchwork of green cells containing normal plastids and white (or yellow) cells containing defective plastids (see Fig. 3.6). (For further details see the extensive review of Kirk & Tilney-Bassett, 1967). In both *Mirabilis* and *Pelargonium* inheritance of the plastid characteristics is non-Mendelian, with the difference that in *Mirabilis*, transmission is exclusively through the female side, whilst in *Pelargonium* it is biparental (see Table 3.2).

In the early work on *Mirabilis* and *Pelargonium* there was no evidence of segregation of nuclear genes controlling variegation.

In more recent work with *Pelargonium*, Tilney-Bassett (1970) confirmed the biparental inheritance of variegation, and showed some additional peculiarities of plastid inheritance. The major control was from the nuclear and plastid genotypes of the female parent, the male having only a minor modifying influence even when plastids from the male parent were transmitted more successfully than those from the female parent. It was also shown that following some crosses, the proportion of variegated embryos was very low, suggesting that there could not have been a random mixing of large numbers of green and white plasmids in the zygote, which probably develops from a cell containing only a few plastids.

Oenothera is another plant which has been the subject of extensive investigations of cytoplasmic genetics (Renner, 1937). Some species of this

Fig. 3.6 An example of a variegated strain of *Pelargonium*. (By kind permission of R. A. E. Tilney-Basset.)

genus have the technical advantage that all the chromosomes become associated in a ring at meiosis, due to a complex system of interchanges, and each of the two parental sets of chromosomes usually segregate as a block. Hence there is little recombination between parental genes and the cytoplasmic genes can be studied against a nearly constant background of blocks of nuclear genes.

Renner showed that in some interspecies crosses, the plastids do not develop normally due to an unfavourable interaction between plastids from one species and nuclear genes from another. Thus, crosses between *Oe. muricata* ♀ and *Oe. hookeri* ♂ produced green F_1 plants, with some yellow sectors, while the reciprocal cross gave only yellow plants which died. It was assumed that the *hookeri* plastids could not develop normally in the presence of some nuclear genes from *Oe. muricata*. Further breeding studies showed that the 'defective' plastids could give rise to normal green plastids when restored to the *hookeri* nuclear genome, even after a long period of association with the unfavourable *muricata* nuclear genome.

Stubbe (1964) and, more recently, Schötz (1974) have classified various species of *Oenothera* into a number of groups, based on compatibility between nuclear and plastid genomes. At least five types of plastid genome have been identified.

Table 3.2 Comparison of inheritance of variegation in *Mirabilis* and *Pelargonium*

		Mirabilis				*Pelargonium*	
	♀	♂	♀	♂	♀	♂	♀	♂
Parents	Green × Variegated		Variegated × Green		Green × Variegated		Variegated × Green
F₁	All green		Green, variegated and white plants		Mainly green, some variegated, some white		

In the past there has been considerable disagreement about the exact location of the extranuclear determinants of plastid characters. Essentially there were two points of view: the plastom hypothesis, according to which the determinants are inside the plastids themselves, and the plasmon hypothesis, which held that the determining factors are located elsewhere in the cytoplasm. It has been thought that if a cell contained two phenotypically different plastids, this would prove the correctness of the plastom hypothesis, and a number of examples of such 'mixed' cells were described. However, such a diversity of plastids in a cell could be due to transitional alterations in chloroplast structure. Therefore, the demonstration that one cell contains two cytologically distinguishable types of chloroplast does not prove one hypothesis or the other. Although much is known about the inheritance of plastid characteristics in higher plants, and in particular about interactions between plastids and nuclear genes, this information is based largely on the occurrence of non-Mendelian behaviour of defective plastids. The chemical basis of these defects is largely unknown. While it is probable that many of the plastid determinants which have been described are controlled by chloroplast genes, definitive proof is still lacking.

Chlamydomonas The unicellular alga *Chlamydomonas reinhardi* has many advantages over higher plants for the study of both nuclear and extranuclear genetics. The vegetative cells contain a single, usually haploid, nucleus, a single chloroplast and some twenty mitochondria (see Fig. 3.7). The organism can be grown in defined media, either in light or darkness, and handled by standard microbiological techniques. The life cycle comprises an asexual phase with a series of mitotic cell divisions, and a sexual phase during which two morphologically identical gametes of opposite mating types (mt^+ and mt^-) fuse together. When the zygote is formed there is fusion both of the two haploid nuclei, and also of the two chloroplasts (see Fig. 3.8). On germination of the zygote, meiosis usually takes place immediately and there is therefore a 2:2 segregation of nuclear genes. Many nuclear genes are known, and are mapped in sixteen linkage groups. One pair of nuclear genes is responsible for mating-type determination.

In 1954 a series of investigations on extranuclear genetics of *Chlamydomonas* was begun by Ruth Sager with the isolation of a mutant showing resistance to a high level of streptomycin (Sager, 1972). This mutant (denoted *sr-2*, later *sm-2*) shows non-Mendelian inheritance. Zygotes produced by crossing streptomycin-resistant with wild type cells give rise to progeny which vary according to the mating-type of the streptomycin-resistant parent: if it is mt^+, nearly all the progeny (usually >99%) are also streptomycin-resistant; if the streptomycin-resistant parent is mt^-, nearly all the progeny are streptomycin-sensitive. No regular, i.e. 2:2, segregation of streptomycin-resistant and sensitive genes occurs during meiosis.

Notwithstanding the predominantly uniparental inheritance of *sm-2*, a

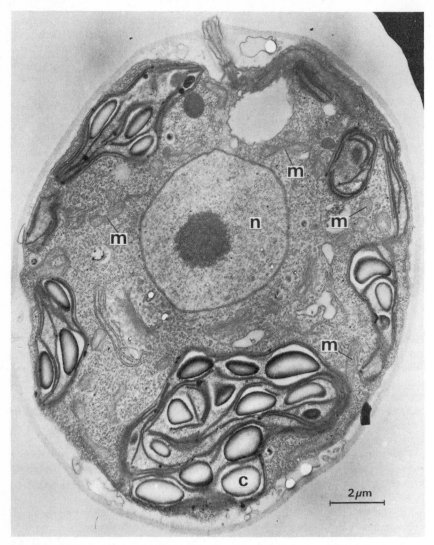

Fig. 3.7 Electron micrograph of a section through a *Chlamydomonas* cell. Chloroplast (c), mitochondria (m) and nucleus (n) are visible. (By kind permission of T. Cavalier-Smith.)

small percentage of zygotes, usually $< 1\%$, can also inherit the *sm-2* factor from the mt^- parent. These are called 'exceptional zygotes'. They usually transmit extranuclear genes from both mating cells, and are therefore 'heterozygotes'. In addition there is a rare class of exceptional zygotes which transmit extranuclear genes *only* from the mt^- parent. The frequency of exceptional zygotes of both types can be raised—sometimes almost to

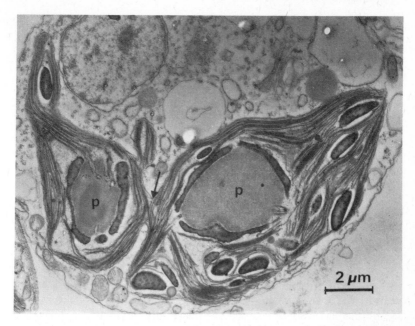

Fig. 3.8 Electron micrograph of a *Chlamydomonas* zygote 4 h 40 min after fusion. It shows the presence of two pyrenoids (p) within a single chloroplast envelope, strongly suggesting the physical fusion (arrow) of the chloroplasts during conjugation. (With permission from T. Cavalier-Smith (1970), *Nature*, **228**, 333–5 and Macmillan, London.

100% — by ultraviolet irradiation of the gametes just before mating, and by some other treatments.

The inheritance of *sm-2* may therefore be described as extranuclear and uniparental (usually via the mt^+ gamete only) with the exceptions mentioned. In later work a large number of other mutants, some showing resistance to different antibiotics, others showing dependence on acetate in the medium, and others temperature-sensitivity, have been described, and show a pattern of inheritance like *sm-2* (Sager, 1972; Gillham, 1974; Sager & Ramanis, 1976).

It should be added that streptomycin-resistance in *Chlamydomonas* may also arise as a consequence of mutation of nuclear genes. Such genes can of course be transmitted to the zygote from gametes of either mating-type, and show a 2:2 segregation at the meiotic stages. A third class of mutations, denoted 'minutes', are extranuclear but regularly show biparental inheritance. They are discussed later (pp. 56, 57).

Streptomycin resistance (*sm-2*) and other variants showing a similar pattern of inheritance, are considered to be due to genes located in the chloroplasts. The evidence for this view, though suggestive, and accepted by the majority of workers in this field, is at present not totally convincing, and the counter-suggestion of control by mitochondrial genes has also been made (Schimmer & Arnold, 1970).

Evidence favouring the chloroplast gene hypothesis has been provided by some experiments of Sager and Lane (1972 who labelled the DNA of gametes of one or other of the mating types with the isotope ^{15}N, and made crosses between cells containing ^{15}N with others containing the normal ^{14}N. It was found that chloroplast DNA in the zygote, six hours after mating contained only ^{14}N if the ^{15}N-containing gamete had been mt^-, but contained ^{15}N if the ^{15}N-containing gamete had been mt^+. Hence, DNA from the mt^+ parent is used in formation of the chloroplast DNA, while chloroplast DNA from the mt^- parent is lost, possibly due to the action of a restriction endonuclease produced by the mt^+ cell, according to a suggestion of Sager and Ramanis (1973), or by some other mechanism, according to Gillham et al. (1974). These experiments show that chloroplast DNA has the same uniparental pattern of inheritance as the gene *sm-2*, and other genes behaving similarly. It should be added, however, that some other studies by Chiang (1968), in which the DNA of one type of gamete was labelled with ^{3}H-adenine, and the other with ^{14}C-adenine, seemed to indicate that the zygote chloroplast contains *both* kinds of DNA, though this result could have been due to some technical problems connected with the use of a 'cold chase' of unlabelled adenine to remove the residue of labelled material in the metabolic pool (Gillham, 1974).

Some other evidence favouring the hypothesis of chloroplast-located genes was obtained by Lee and Jones (1973), who treated *Chlamydomonas* cells with a chemical mutagen at the time of replication of chloroplast DNA, and produced a small $(1.6 \times)$ increase in the frequency of extranuclear mutations. However, the same treatment produced a similar rise in the number of *nuclear* mutations; and in any case the time of replication of mitochondrial DNA in these experiments was not ascertained. Hence the results of this experiment are rather inconclusive.

As for the phenotypic effect of the various extranuclear drug-resistant mutations, it is true that some of them affect the chloroplast ribosomes (see below p. 60). Moreover some extrachromosomal mutants show loss of photosynthetic capacity, i.e. affect the chloroplasts. However, the site of the phenotypic expression of a mutant gene tells us nothing about the site of the coding of the genetic information concerned. In fact, nuclear genes are also known to affect the chloroplast ribosomes of *Chlamydomonas* (see p. 60), while a number of drug-resistant extranuclear mutants of *Chlamydomonas* affect both chloroplasts and mitochondria (Gillham, 1974).

As mentioned above, there is a class of extranuclear mutants of *Chlamydomonas*, known as 'minutes', which regularly show biparental inheritance (Alexander et al., 1974). These can be induced by treatment with acriflavine or ethidium bromide, substances known to be highly effective inducers of petite, i.e. mitochondrial, mutants in yeast. In *Chlamydomonas* the minute mutants live for 8 to 9 cell divisions in the light, and then die. They can be mated with normal cells during the interval between induction and death, and the zygotes obtained show a segregation of 4:0, or 0:4, mutants:normals. Hence they cannot be due to nuclear genes. The minute mutants, unlike *sm-2*, can be transmitted equally readily throughhe mt^+ or mt^- gametes, and therefore may be said to show an extranuclear but biparental pattern of inheritance. Recently, another class of mutants denoted 'dark-diers' (*dk*), having a mode of biparental inheritance like the minutes, has been described (Wiseman *et al.*, 1977). In view of a number of similarities with the petite yeasts, the *Chlamydomonas* minutes and 'dark-diers' are considered to be controlled by mitochondrial genes. This is not strictly relevant to the question of the location of the genes for the uniparentally inherited factors like *sm-2*, but at least one can say that there are probably two distinct classes of extranuclear genes in *Chlamydomonas*, and if one class is based on mitochondrial DNA, the other is probably based on chloroplast DNA.

We have tried to summarize rather critically the evidence for the existence of a system of chloroplast genes in *Chlamydomonas*. Notwithstanding the apparent lack of much direct supporting evidence, we will proceed on the assumption that such a system is in fact responsible for the mutants considered here.

A large quantity of data has been collected on the inheritance of various chloroplast genes in *Chlamydomonas* (Sager, 1972; Gillham, 1974; Gillham *et al.*, 1974; Sager & Ramanis, 1976), though different interpretations of some of the results have been put forward by different workers. These conflicting interpretations seem to be due in part to lack of agreement on certain basic facts, for example, as to whether the chloroplast contains only two or a larger number of DNA strands or chromosomes.

It seems clear that segregation and recombination of chloroplast genes occur in the progeny of the 'exceptional' zygotes as a result of events taking place both in the zygote cell itself (i.e. during the two meiotic divisions) and more especially during the succeeding mitotic divisions. This continues for some ten postmeiotic divisions, after which nearly all the cells are homozygous in regard to all chloroplast genes. By carefully following the progeny of individual cells at various stages, Sager (1972) found that segregation was of two types, denoted types II and III. Type II segregation is the production by a given heterozygous cell of two daughters, one of which is again heterozygous while the other is homozygous for one or other of a pair of alleles; type III segregation is the production by a heterozygous cell of two daughters, one homozygous for one allele, the other homozygous for the other allele. Type II segregation is considered to be due to a process resembling gene conversion, while type III segregation is considered to involve two reciprocal recombination events near to the locus concerned. The frequency of type II segregation is approximately the same for all chloroplast genes studied, about 39% over the period of the first and second mitoses after zygote germination; the frequency of type III segregation, however, varies according to the locus, from 6% for *ac* to 12% for *ery* (Sager & Ramanis, 1976).

According to the interpretation of Sager, type III segregation results from a process resembling mitotic crossing-over in the nuclear chromosomes of eukaryotic organisms (Pontecorvo, 1958), and it is assumed that each chloroplast has two chromosomes each containing a centromere-like attachment point. (This is apparently the first time that a centromere has been postulated to occur in prokaryotic or organelle DNA.) Genes which are more remote from the attachment point would be expected to show a higher frequency of type III segregation than genes close to the attachment point. Thus, according to this view, all type III segregation involves recombination.

A second method of map construction is based on 'co-segregation' of two genes. The assumption here is that two genes which are close together will show more frequent co-segregation than two which are far apart. A third method, which appears to be an extension of the first, is based on estimation of the rate at which heterozygotes give rise to homozygotes over a period of successive mitotic divisions: the more rapid this is, the farther a given mutant gene should be from the attachment point. Taking together numerous data analysed by these methods, Sager and Ramanis (1976) constructed a circular map of the chloroplast genome of *Chlamydomonas*.

Since recombination occurs frequently during the course of the cell divisions following germination of a zygote heterozygous for a number of chloroplast genes, gene mapping by taking the end-result of all this recombination and simply estimating the relative proportions of recombinants and non-recombinants, is impracticable, except for closely linked markers. Unfortunately, the alternative methods based on deletion mapping and other techniques which are available in work with mitochondrial genes in yeast, cannot be used in studies on chloroplast genes of *Chlamydomonas*. A detailed critical discussion of mapping of chloroplast genes in *Chlamydomonas* is given by Adams *et al.*, 1976.

The dual genetic control of chloroplasts

Introduction   Chloroplasts, like mitochondria, require the activity of two sets of genes, one in the cell nucleus and one in the organelle, for the correct development of chloroplast function. As in Chapter 2 on mitochondria, we aim here to discuss the functions of both sets of genes but with particular emphasis on the function of organelle genes. Chloroplast DNA is larger than mitochondrial DNA, about eight times larger than the mitochondrial DNA of mammals. This might suggest that the role of the chloroplast DNA in chloroplast biogenesis is more complex than the role of mitochondrial DNA in mitochondrial biogenesis. Chloroplast DNA could, in principle, determine the structure of between 100 and 150 proteins.

The methods used to study the products of the chloroplast genome are essentially similar to those used to investigate the products of the mitochondrial genome. There are essentially two such methods. Firstly, the hybridization of some chloroplast RNAs to chloroplast DNA shows that these RNAs are complementary to sequences on the chloroplast DNA. This technique has not as yet been exploited as fully with chloroplasts as with mitochondria, perhaps due to the absence in chloroplasts of the deletion mutants that exist in mitochondrial petite mutants of yeast. However, the study of the translation of chloroplast messenger RNAs has been more successful than that of translation of mitochondrial messenger RNAs.

The second technique is that of standard genetic analysis. If a heritable change in the chloroplast DNA can be correlated with a primary change in a chloroplast component such as a protein, this implies that the component is determined by the chloroplast DNA. This technique has proved useful for the analysis of chloroplast genes coding for ribosomal proteins in *Chlamydomonas* and fraction I protein in *Nicotiana*.

Apart from these two methods, much work has been done using specific inhibitors of chloroplast transcription and translation. These are, however, subject to the same limitations as those outlined in Chapter 2 for inhibitor studies on mitochondria. Clearer results come from the study of proteins synthesized by isolated chloroplasts. Ellis and his co-workers have developed a system in which intact chloroplasts are able to synthesize polypeptide chains *in vitro* using energy derived from light and an artificial ATP generating system (Ellis, 1975).

Studies of the greening process, whereby an etioplast becomes a functional chloroplast, have revealed that a large number of nuclear genes in both algae and higher plants are involved in the production of a functioning photosynthetic system. Chloroplast differentiation is a very complex process and many mutants are pleiotropic in their effects. (For more details, see Kirk & Tilney-Bassett, 1967 and Boardman *et al.*, 1971).

A good example of this type of study is the collection of nuclear gene mutations in barley affecting chloroplast formation (von Wettstein *et al.*, 1971). There are 86 different nuclear gene loci in barley, each influencing a specific aspect of chloroplast development.

We now propose to review the various functional systems of the chloroplast with particular reference to the location of the genes, nuclear or chloroplast, which determine the components of these systems.

Replication and transcription of chloroplast DNA The manner in which chloroplast DNA is replicated has been discussed briefly in the first section of this chapter but there is as yet no evidence regarding the origin of the replication enzymes. Chloroplasts have been shown to incorporate ^{3}H-uridine into RNA in the presence of light. Several RNA classes are labelled, the most highly labelled being chloroplast ribosomal RNA. The chloroplast RNA polymerase, in contrast to nuclear RNA polymerase, is sensitive to the antibiotic rifampicin (Loiseaux *et al.*, 1975). However, as the chloroplast polymerase persists in cells after prolonged bleaching treatment, it is probable that this enzyme is nuclearly coded, although this has not been proved directly (McLennan & Keir, 1975).

Chloroplast ribosomes and protein synthesis As stated earlier there are two large chloroplast ribosomal RNAs. It seems that these RNAs are transcribed from chloroplast DNA. In particular Stutz (1971) found that the heavy strand of chloroplast DNA bound nine times more chloroplast ribosomal RNA than the light strand. While the significance of this result is not clear it is interesting to note that Attardi *et al.* (1976) obtained substantially the same results in studies on mitochondrial nucleic acids from HeLa cells.

Tewari and Wildman (1970) have shown that between 20 and 30 t-RNAs are probably coded by the chloroplast DNA of *Nicotiana*. Recent evidence suggests that *Euglena* chloroplast DNA may also code for about 26 t-RNAs (Schwartzbach *et al.*, 1976).

In *Chlamydomonas*, many mutants resistant to antibacterial antibiotics have been described, and most of these mutants produce changes in the chloroplast ribosome, since ribosomes isolated from chloroplasts of resistant cells are able to synthesize polypeptides *in vitro* in the presence of antibiotics. This situation is in some senses similar to that of the antibiotic resistant mutants described in the mitochondria of several organisms. While the position is not yet completely clear, there is good evidence that in chloroplasts it is the chloroplast ribosomal proteins which are modified in the antibiotic resistant stocks. Bogorad and co-workers have made a study of three different types of mutant giving erythromycin-resistant chloroplast ribosomes in *Chlamydomonas reinhardi* (Bogorad, 1975).

Genetic studies showed that two of these mutant classes behaved at meiosis as if due to nuclear genes. A third class behaved in a non-Mendelian manner in that erythromycin-resistance was largely maternally inherited, like the chloroplast genes studied by Sager. This suggests that this mutant class is determined by chloroplast DNA. Further analysis of the chloroplast proteins by two-dimensional gel electrophoretic techniques showed that each mutant type modified one chloroplast ribosomal protein. The importance of Bogorad's finding for this chapter is the demonstration that both the nuclear and chloroplast genomes are required to produce functional chloroplast ribosomes.

Fraction I protein and other stromal proteins The soluble proteins of the chloroplast are found largely in the stroma surrounding the lamellae. Nearly 40% of this class of protein in higher plant chloroplasts consists of one single protein species. This species is fraction I protein, the enzyme responsible for the first step of the fixation of CO_2. It has a molecular weight of about 5.2×10^5 and is composed of two types of subunit, one large (5.5×10^4) and one small (1.4×10^4). Because of its abundance in the chloroplast, this protein has naturally been the subject of extensive study. Working with *Nicotiana* species, Kawashima and Wildman (1972) found interspecies differences in the peptides produced by tryptic digests of the small subunit of fraction I protein. In crosses between different species designed to follow the inheritance of tryptic peptides, it was shown that the small subunit was coded by nuclear genes. In this study no variation was found between the tryptic peptides of the large subunits.

A later study by Chan and Wildman (1972) involving a comparison between species from both the eastern and western hemispheres showed that the large subunit isolated from Australian species of *Nicotiana* contained one tryptic peptide not found in the species of the western hemisphere. By crossing the western species with the Australian species it was shown that the large subunit, unlike the small subunit, was inherited maternally and therefore was coded by extranuclear genes probably located in the chloroplast. Subsequently it has been shown that isolated chloroplasts do indeed synthesize the large subunit *in vitro* (Blair & Ellis, 1973). Hartley, Wheeler and Ellis (1975) have been able to isolate the messenger RNA for the large subunit from chloroplasts and obtain *in vitro* translation of a recognizable polypeptide in a cell-free *E. coli* system. Recently, Gelvin *et al.* (1977), working with *Chlamydomonas*, isolated the large subunit m-RNA, which was hybridized with a fragment of chloroplast DNA. Thus the coding of the larger subunit by a chloroplast gene was confirmed.

In summary, fraction I protein, like the chloroplast ribosomes, is a product of the cooperation of both nuclear and chloroplast genes. The small subunit is determined by nuclear genes and synthesized on cytoplasmic ribosomes. It is then transported across the chloroplast membrane where it forms the functional enzyme by association with the large subunit. This is coded by a chloroplast gene and synthesized on chloroplast ribosomes.

Membrane proteins Much of the chloroplast-synthesized protein apart from fraction I protein is found associated with the various membrane systems of the chloroplast (see review by Ellis, 1976). The clear genetic evidence for the chloroplast determination of the ribosomal proteins and fraction I protein large subunit is not available for other chloroplast proteins at the moment. However it is worthwhile to consider other proteins synthesized by isolated chloroplasts as it is likely that most if not all of these proteins will turn out to be coded by chloroplast DNA. Recently Ellis and co-workers (see Ellis 1975) have found that isolated chloroplasts synthesized about thirty discrete proteins, three of which have been identified as subunits of the chloroplast coupling factor (CF_1) (part of the chloroplast ATP-ase). This contrasts with the situation in mitochondria where all the ATP-ase F_1 subunits are made on cytoplasmic, rather than mitochondrial, ribosomes. Hence it is likely that chloroplast DNA codes for a wider range of products than does mitochondrial DNA. There is little information on the function of other chloroplast-synthesized components. This is in large part due to our lack of knowledge about the proteins present in chloroplast membranes.

It seems that chloroplast-synthesized proteins are apparently not closely associated with either of the two photosystems or with cytochrome f, these being two components at present well characterized. In another study of protein synthesis in isolated chloroplasts, Joy and Ellis (see Ellis, 1975) concluded that two or possibly three of the polypeptides of the chloroplast envelope are synthesized on chloroplast ribosomes. There is, as yet, no evidence as to the function of these proteins.

It is probable that the number of proteins synthesized by chloroplast ribosomes may be small, perhaps as few as 30. One or more of them are ribosomal proteins associated with the site of ribosomal erythromycin resistance and three are associated with the chloroplast ATP-ase. Others are associated with chloroplast membranes, both internal and external and these may be associated with the complex enzyme systems responsible for light driven electron transport. The chloroplast ribosomes also synthesize the large subunit of fraction I protein. The evidence that chloroplast ribosomes make only a limited number of proteins implies that the majority of proteins are synthesized on cytoplasmic ribosomes and transported across the chloroplast membrane. A summary of our current knowledge of the contribution of the nuclear and chloroplast genetic systems to chloroplast biogenesis is shown in Table 3.3.

Conclusion

At a very basic level, the function of the chloroplast genome is similar to the function of the mitochondrial genome (compare Tables 3.3 and 2.4). This function is to provide ribosomal RNAs and a few other components of the organelle protein synthetic system to translate a few messenger RNAs probably coded by the organelle DNA. These proteins are largely associated with the chloroplast membranes where they may form part of complex hydrophobic enzymes. The majority of the proteins found in both chloroplasts and mitochondria are determined by nuclear genes and synthesized on cytoplasmic ribosomes.

There are, however, many differences between the two organelles. Perhaps the most obvious is that the major product of chloroplast protein synthesis, fraction I protein large subunit, is a soluble protein which is the most abundant protein species in the plant. By contrast, mitochondrial protein products are a very small proportion of the total protein of a cell. In addition the structural complexity of chloroplast membranes appears greater than that of the mitochondrial cristae. The potential coding capacity of chloroplast DNA is also substantially larger than that of mitochondrial DNA.

It must be said that we are only beginning to understand the complexity of chloroplast biogenesis; our knowledge of many areas is far from complete. It is very likely that as the recent technical advances of molecular genetics and biochemistry are applied to the study of organelles, our ideas about the function of the chloroplast genome will change.

Table 3.3 Genetic control of chloroplast constituents (provisional summary)

Constituent	Controlled by nuclear genome	Controlled by chloroplast genome
1 Chloroplast DNA polynucleotide structure	—	✓
2 Replication, recombination transcription of chloroplast DNA	✓	?
3 Protein synthesing system of chloroplast		
r-RNA	—	✓
ribosomal proteins	✓	✓
t-RNA	✓	✓
4 Soluble enzymes	✓	?
5 Fraction one protein		
large subunit	—	✓
small subunit	✓	—
6 Lamellae and photosystems I and II	✓	✓
7 Chloroplast envelope	✓	✓

N.B. Both nuclear and chloroplast mutations are known to affect the assembly of the lamellae and photosystems but most are pleiotropic and the primary effect is unknown.

4 Plasmids

Introduction

In bacteria the genetic system controlling the overwhelming majority of hereditary characters consists basically of a circular structure made of DNA. This is conventionally called the bacterial 'chromosome', even though it lacks some of the features of true eukaryotic chromosomes. Unlike them, the bacterial chromosome does not contain basic proteins, has no centromere, is not distributed between dividing cells by mitosis and does not undergo meiosis. The bacterial chromosome is also, of course, much smaller than 'real' eukaryotic chromosomes. Nevertheless, it is convenient to use the word 'chromosome' for the main mass of bacterial DNA. In addition to this chromosomal material, some bacteria contain small accessory DNA circles, bearing genes controlling a small proportion of the heredity of the bacteria. The accessory strands are called 'plasmids', a word originally used by Lederberg (1952) for any extrachromosomal hereditary determinant, but now usually taken to have a more restricted meaning.

Plasmids are certainly extranuclear genetic components, but the parallel with non-Mendelian systems in higher organisms is not particularly close. In higher organisms there is a Mendelian system, based on the chromosomes in the nucleus, and one or more non-Mendelian systems, based on DNA-containing structures or organelles in the cytoplasm; and there is a definite partitioning of the cell into nucleus and cytoplasm. In bacteria even the main chromosomal system can hardly be called 'Mendelian', and nucleus and cytoplasm are not separated by a membrane. However, in spite of the weak parallelism between cytoplasmic organelles in higher organisms and plasmids in bacteria, plasmids cannot possibly be excluded from a survey of extranuclear genetic systems. Plasmids have been very intensively studied by the most up-to-date techniques of molecular biology, and the amount of detailed knowledge about them is greater than for the other cellular components discussed in this book. Furthermore, in a number of ways, plasmids can be taken as models for comparison with other structures, and suggest the types of investigation which should be carried out.

Needless to say, no attempt is made here to cover the ever-growing literature on plasmids. This account will contain only brief discussions of some well-known types, a description of their more striking characteristics, and of the relative roles of the two genetic systems, the one in the bacterial chromosomes and the other in the plasmids. Fuller accounts will be found in various reviews (Clowes, 1972; Falkow, 1975).

It is rather difficult to define the word plasmid precisely, because features most characteristic of some plasmids are absent in others, and some of these characteristics are exhibited by other particles, such as viruses, which are not usually classified as plasmids. Perhaps the only firm statements which can be made are that plasmids are genetic structures, are made of DNA, are smaller than the bacterial chromosome, are spatially separate from the chromosome and are capable of replication. They are often described as 'autonomous' in their reproduction, but this is only partly true in view of the control over their replication by the bacterial chromosomal genome. It is true that the specific molecular structure, or nucleotide sequence, of plasmid DNA is determined by that of pre-exising, parental, plasmid DNA, and in this plasmids may be said to show autonomous replication; but plasmids are not peculiar in this, since it applies to many other DNA-containing structures. Other properties, such as control of bacterial conjugation, reversible insertion in bacterial chromosomes and transferability from one bacterium to another—which will be outlined in the following pages—apply to some plasmids but not to others, and so cannot be included in a general definition.

In this chapter bacteriophages are arbitrarily excluded from consideration as plasmids. Phages have some properties, such as unrestricted replication and development of a number of structural and enzymatic proteins, not found in most plasmids; though some phages (e.g. phage λ in *E. coli*) and plasmids have a number of features in common. Some writers group them together.

Brief description of F, R and Col plasmids

The F or sex factor   The sex or fertility factors of *E. coli* were discovered in the course of early work on the genetics of these bacteria (Hayes, 1968). It was found that recombination of bacterial characters occurs following a mixing together of two different strains, denoted F+ and F−. The former is a kind of 'male' or donor of genetic material, and the latter a kind of 'female' or recipient. It is now known that two quite different processes may take place when F+ and F− cells come into contact. In the first, the F factor itself passes from the F+ to the F− bacterium, converting the latter to the F+ (donor) state, but no transfer of genes located in the bacterial chromosome takes place. In the second process, part—or rarely, the whole—of the bacterial chromosome of the F+ bacterium is transferred to the F− bacterium, but the latter does not usually acquire the ability to act as a donor of genetic material (for further details, see below p. 71). The evidence for these statements will be found in textbooks of bacterial genetics (Hayes, 1968; Stent, 1971) and reviews (Curtiss, 1969; Clowes, 1972; Reeve & Willetts, 1974). Detailed studies established that the first type of process takes place when an F factor is separate from the bacterial chromosome. In this situation the F factor is a kind of parasite which is able to transfer itself from one bacterium to another and confers no obvious benefit on the host bacterium. For the second type of process to occur, however, it is first necessary for the F factor to become inserted within the bacterial chromosome at one of a number of specific sites. Bacteria in this state are denoted Hfr, meaning high frequency of recombination. The inserted F factor is able to propel the bacterial chromosome into the F− bacterium, where recombination of bacterial genes may subsequently take place. Usually the process is interrupted before the whole of the donor cell chromosome is transferred. Moreover, that part of the F factor concerned with the transfer process (the group of *tra* elements, see p. 71) is located at the distal end of the transferred chromosome and is usually left behind in the donor cell.

The most conspicuous property of F is therefore to initiate conjugation of bacteria and transfer either itself, or part, or rarely all, of the bacterial chromosomal genes from one bacterium to another. Having the capacity of existing in either of two alternative situations, the one 'free', the other inserted in the bacterial chromosome, F has been included in the category of cell constituents denoted 'episomes' (Jacob & Wollman, 1958). As will be described, however, not all plasmids have episomal properties: they do not all have the ability to insert themselves in bacterial chromosomes. Moreover, not all plasmids instigate bacterial conjugation.

F is a plasmid because it is sometimes located outside the bacterial chromosome and because it is a small circular DNA-containing structure. Further details of F plasmids will be described later in this Chapter.

The R or resistance transfer factors   R factors cause bacteria containing them to show resistance to concentrations of antibiotics which kill ordinary sensitive bacteria. In 1956 the alarming discovery was made in Japan that resistance of *Shigella* to a number of antibiotics, such as streptomycin, tetracycline, chloramphenicol and the sulphonamides, could arise in a single step (Watanabe, 1963; Falkow, 1975). It was shown that this was due to the simultaneous transmission of several resistance genes from another bacterium. Such transmission could occur even between bacteria belonging to widely separated species or genera such as *Escherichia*, *Shigella* and *Proteus*. The R factor thus not only determines resistance to antibiotics but also enables bacteria possessing it to conjugate and thereby transfer the factor to other bacteria.

In some respects R resembles F, but the two factors are distinct since transfer of resistance genes may take place between bacteria which both lack F. R factors do not have the capacity of becoming inserted in the bacterial chromosome in the manner described for F, whereby Hfr bacteria are formed. However, some R factors, in a way which does not seem to be clearly understood, do have the ability to 'mobilize' bacterial genes, at a low frequency, and thus occasionally transfer them to other bacteria (Meynell, 1972).

The R factor, like F, has been shown to consist of a small extrachromosomal circle of DNA (Clowes, 1972). It is therefore a plasmid. As will be described later, some R plasmids consist of two parts, one bearing genes for antibiotic resistance, the other bearing genes making possible the transfer of the plasmid to another bacterium.

Some R factors, like those causing penicillin-resistance in *Staphylococcus*, do not have the ability to transfer themselves from one bacterium to another. They contain genes for antibiotic resistance but not for transfer. The extrachromosomal location of these factors was inferred from other observations, such as their capacity to be irreversibly lost from the bacteria. Such plasmids can however be transferred from one bacterium to another by transduction (Richmond, 1968; Lacey, 1975).

Col factors Colicins, which were studied for many years before their genetic basis was understood (Frédéricq, 1957; 1963; Hardy, 1975), are toxic proteins which are produced by bacteria, such as *Escherichia*, *Shigella* or *Salmonella*, and which kill other bacteria. The colicin-producing bacteria are immune to the lethal action of the specific type of colicin produced by themselves. Colicin production is a stable hereditary characteristic. In part the phenomenon resembles lysogenesis, whereby a bacterium liberates a phage which is capable of killing other bacteria; but colicins differ from phages in that, after sensitive bacteria are killed by colicins, no new generation of colicin factors is liberated, whereas after lysis of bacteria by phage, a new burst of phage particles is produced. Colicinogenic bacteria carry the potentiality to synthesize colicin, but this is normally expressed by only a small proportion of cells in a population, which die in the process. Induction of this expression can be brought about by ultra-violet light and other agents—another point of resemblance to lysogenesis.

Genetic studies have shown that colicin production is not determined by genes in the bacterial chromosome. The determinants can be transmitted from one bacterium to another in a similar manner to that of the F plasmids, but the detailed mechanism is different. Transfer of colicin determinants may take place between two bacteria both of which are F − ; hence F plasmids need not be involved. Transmission of the ability to produce colicins, when it occurs, is usually at a lower rate than transmission of F, and the bacterial structures mediating the transfer of the two factors (sex pili, see p. 71) may be different (Meynell & Lawn, 1967). Although some colicin factors are like F able to mediate their own transfer, others (e.g. Col E1) are unable to do this, and require the aid of a sex-factor, such as F, for such transfer.

Some colicin factors have the ability to 'mobilize' the transfer of bacterial chromosomes. Those that do fall into two classes: (1) those that become inserted in the bacterial chromosome and form Hfr types (Kahn, 1968), and (2) those that mediate chromosome transfer in some other way, not involving permanent insertion in the bacterial chromosomes (Smith & Stocker, 1966), like the R-mediated transfer mentioned above.

Classification of plasmids The three types of plasmid described above—
F, R, Col—are named by their most conspicuous effects: fertility, or ability
to transfer genetic material between bacteria; drug-resistance, and colicin
production. However, some plasmids, denoted in this way, exhibit more
than one of the three effects, even though the nomenclature indicates only
one of them; and some plasmids have effects other than the three described,
e.g. some confer pathogenicity on the host bacteria (Meynell, 1972; Falkow,
1975). Specification of plasmids in terms of their main effects is therefore an
unsatisfactory method of classification, and Datta (1975) proposed
classifying plasmids into groups based on 'compatibility', i.e. ability to
exclude other plasmids of the same type from coexistence in the same
bacterium. Nevertheless, naturally occurring plasmids continue to be
denoted by one of the three symbols F, R, Col. (For a recent discussion of
the nomenclature of plasmids, see Novick *et al.*, 1976.)

Intensive work is going on to characterize individual plasmids more
completely, and in a few cases substantial portions of the entire plasmid
genome have been mapped, as will be indicated later (see p. 75). Such work
will, of course, facilitate classification, or possibly show that it is impossible
to separate plasmids into taxonomic groups like those in which more
complete organisms are organized.

Plasmid DNA and its replication

All plasmids so far identified consist of small DNA circles, which may
be either single or multiple structures linked together in chains. Sometimes 'supercoils', i.e. covalently closed duplex molecules twisted upon
themselves, are seen, resembling the DNA of mitochondria (see p. 14). The
size of the DNA circles varies greatly in different plasmids. The F plasmid of
E. coli is about 30 µm in length, and has a molecular weight of 62×10^6
daltons (Achtman, 1973). The size is approximately two per cent of that of
the *E. coli* chromosome. The size of R plasmids is usually similar to F, but
other plasmids vary from the minute P16C of *E. coli*, which is 0.05% of the
chromosome (Ikeda *et al.*, 1970), to the relatively enormous compound
plasmids, such as some F prime (F') types, though these include some
segments of bacterial DNA. Colicin plasmids such as ColE1, E2 or E3,
which do not contain transfer genes, are relatively small, with a molecular
weight of about 5×10^6 daltons, and a length of 2.3 µm (Roth & Helinski,
1967).

The number of plasmids in a single bacterium varies considerably. F and
R plasmids usually occur as one or only a few per bacterial chromosome, but
the smaller plasmids may be more numerous. Cabello *et al.* (1976) estimate
the number of 18 for ColE1, and Novick *et al.* (1975) estimate 20–30 for the
smaller *Staphylococcus* plasmids.

The base composition (GC content) of plasmid DNA may be
indistinguishable from that of the host bacterial chromosome. This is true of
F in *E. coli*, which has a GC content of c. 50%. However, when F is

transferred to other bacteria, whose DNAs have substantially different base compositions, e.g. *Serratia marcescens* (GC 58%) or *Proteus mirabilis* (GC 40%), the plasmid DNA (GC 50%) can be seen to be quite distinct (Falkow *et al.*, 1964).

Since plasmids consist of DNA and very little else, their reproduction is in effect the replication of this DNA. In accordance with the Watson-Crick hypothesis, this replication is now understood to consist primarily of a template-line copying of the nucleotide sequences in the duplex poly-nucleotide chains, by a semi-conservative mechanism. However, many factors, some still poorly understood, are involved.

Our thinking on the question of DNA replication has been much influenced by a paper by Jacob, Brenner and Cuzin (1963), in which the replicon hypothesis was proposed. The replicon was defined as a genetic element capable of independent replication. A whole chromosome, whether in a eukaryotic or prokaryotic cell, accords with the definition, but loose fragments of chromosomes or individual genes do not. Plasmids obviously do. Jacob *et al.* developed the replicon hypothesis from studies with the F-lac plasmid of *E. coli*. This is an F plasmid containing, as an additional insertion, a segment of bacterial chromosomal DNA bearing the genes for lactose fermentation. It was shown that when these bacterial genes were inserted in the non-integrated F plasmid, their replication was under the control of the plasmid. On the other hand, when the plasmid was inserted in the bacterial chromosome, replication was coordinated with that of the chromosome. It was therefore proposed that both bacterial chromosome and plasmid, and every other replicon, contains a special region controlling replication of the whole. This control is thought to be exercised by the synthesis of a specific initiator, acting on a unique initiation site. Once replication starts at this point, it is thought to travel in a prescribed direction and time around the circular DNA strand. The details may vary for different types of DNA, for example replication of small plasmids is thought to be unidirectional (Helinski *et al.*, 1975), but for larger replicons such as phage lambda it is bidirectional.

When a plasmid, including its initiation locus, is inserted in a bacterial chromosome, as in Hfr bacteria, the plasmid initiation mechanism is presumed to be 'switched off', due to the overriding control of the bacterial chromosome mechanism. (As a further complication, some plasmids have *two* origins at which replication starts (Perlman & Rownd, 1976; Cabello *et al.*, 1976).)

Apart from initiation, plasmid DNA replication is thought to involve at least two other processes: polymerization and termination. The three processes have been defined as follows: initiation comprises all events preceding insertion of the first nucleotides; polymerization is the synthesis of complementary replica strands, and termination is the conversion of the replicas into covalently closed monomers (Novick *et al.*, 1975). Many other details of DNA replication are under intensive study. Two main schemes have been proposed: (1) the Cairns hypothesis, and (2) the 'rolling-circle' hypothesis (Cairns, 1963; Gilbert & Dressler, 1968). It is generally thought that vegetative reproduction of plasmids follows the Cairns scheme, whereas certain specialized types of replication, such as during conjugal transfer, may involve the rolling circle mechanism.

Having dissected the DNA replication process into initiation, polymerization and termination, it is possible to ask which components are controlled by plasmid genes and which by genes in the bacterial chromosome. Temperature-sensitive mutants affecting replication of plasmid DNA have been used to give information about this. At a permissive (i.e. moderate) temperature, these mutants continue to replicate, but at a restrictive (i.e. high) temperature, they do not and the plasmids are lost. It has been found that mutations in both plasmid and bacterial genes may affect DNA replication, but initiation is plasmid specific, i.e. controlled by plasmid genes only (Helinski *et al.*, 1975; Novick *et al.*, 1975). Polymerization of plasmid DNA, on the other hand, has been shown to be controlled by bacterial genes. For example a bacterial gene—*dna*-E, controlling DNA polymerase III—is known to affect plasmid replication in an R factor in *E. coli* (Nordström *et al.*, 1974). It is provisionally concluded that initiation is the only plasmid-controlled part of the replication process, though even this may be suppressed by the bacterial system under certain circumstances. Most or all of the remaining stages are thought to be under bacterial control (Novick *et al.*, 1975). Future work will show how far this generalization is valid. Obviously plasmid replication must be partly linked to bacterial replication. If it were not, plasmids would sometimes increase *ad lib* and destroy the host cell, as phages sometimes do. In other situations plasmids would fail to reproduce rapidly enough and so would be lost.

The number of plasmids per cell (or more precisely, the number per bacterial chromosome) is also controlled by both plasmid and bacterial genes. Mutants of R factors in *E. coli* are known which increase the number of plasmids per cell twice or four times (Nordström *et al.*, 1972), and the sites of these genes are known to be located on the segment of the plasmid genome controlling drug resistance. Copy number seems to be a property of plasmid replication, and according to a hypothesis of Pritchard *et al.* (1969), may be regulated by a repressor of replication (Cabello *et al.*, 1976). On the other hand, Rownd *et al.* (1966) found that an R plasmid, present as single examples in *E. coli*, formed twelve copies per cell when transferred to *Proteus mirabilis*, indicating *bacterial* control of plasmid number.

When a plasmid is inserted in the bacterial chromosome, bacterial control is absolute, but when the plasmid is 'free', bacterial control is something less than absolute. Plasmids such as F or R which only form one or a small number of copies per bacterium, must be subject to rigorous control in regard to co-ordination of plasmid and chromosomal replication. This is denoted 'stringent control' (Collins & Pritchard, 1973), but it is not absolute. For example, at times of transfer from F+ to F− cells, plasmid DNA replication can occur in the complete absence of chromosomal replication (Wilkins & Hollom, 1974). Plasmids which are relatively numerous (10–30 per cell) are said to be under 'relaxed' control, though even here the number must be regulated within certain limits.

The mechanism controlling plasmid replication has been thought by some authors to include control by a limited number of special 'membrane attachment sites', which were suggested originally by Jacob *et al.* (1963). Some circumstantial evidence in favour of such sites has been described (Clewell & Helinski, 1969), but direct evidence is rather sparse, and some recent findings (Cabello *et al.*, 1976) seem to go against the idea.

To sum up one may say that, notwithstanding many uncertainties, we are beginning to become aware of the various factors involved in the mechanism of plasmid reproduction, especially in regard to its control by plasmid and bacterial genes. This information is of great value in helping us to probe into the reproduction of cellular organelles, symbionts and virus-like particles in cells.

Transfer of plasmids from one bacterium to another

As already mentioned, some plasmids have the remarkable ability of transferring themselves from one bacterium to another. This discovery originally came from studies on genetic recombination in *E. coli*. Nothing comparable has been found in eukaryotic organisms. An essential element in the transfer mechanism is the sex pilus, a hair-like outgrowth from the outer membrane of the donor bacterium. The F sex pilus may be up to 20 μm in length (though it often appears shorter in EM pictures) and about 8 nm wide (Meynell, 1972). The surface of sex pili is receptive to the attachment of certain phages, and this is one method by which the sex pili produced by different plasmids can be distinguished. For exaple ColI plasmids produce a sex pilus described as 'I-like', having a length of only 1.5 μm and susceptible to a type of phage different from that adsorbed by the longer 'F-like' sex pili produced by F plasmids. Some R plasmids produce I-like sex pili, others F-like. Moreover, there are further subdivisions of these two types, and other types exist (Falkow, 1975).

When a sex pilus is removed by mechanical injury, or prevented from being formed by genetic or environmental means, no trasfer of plasmid DNA takes place. However, the precise role of the sex pilus in the transfer process is not known. It has been thought to act as a kind of male organ, but DNA has never been identified inside the pilus. Alternatively, it may act merely to hold together the two mating bacteria. In this connection it is perhaps noteworthy that the sex pilus has been shown to be retractible (Marvin & Hohn, 1969).

The transfer process in *E. coli* is controlled by a group of closely linked plasmid genes, denoted *tra-A*, *-B*, *-C* etc. These genes are concerned with formation of the sex pilus and maintenance of conditions necessary for transfer to take place. Mutations of the *tra* genes, or deletions of the DNA segment containing these genes, result in loss of ability to transfer plasmid DNA. When two plasmids (or parts of plasmids) containing different *tra* mutants are introduced into the same bacterial cell, the defect of one mutant may be made good by the normal elements in the other, and vice versa. Complementaton may occur, and some twelve distinct genes or cistrons have been identified by complementation analysis (Willetts & Achtman, 1972). Eight of these genes are required for pilus production, one being the structural gene coding for the pilus protein; three are required for 'processing' the plasmid DNA in some way necessary for transfer, and one gene regulates the expression of the other eleven.

Prior to transfer, the plasmid is thought to become attached to the inner surface of the bacterial membrane. Due to the action of a gene in the *tra* series, one of the plasmid DNA strands is 'nicked' at a site called the origin of transfer (*ori-T*), and the linear strand thus formed moves across into the recipient bacterium. Thus single plasmid DNA strands are produced in donor and recipient bacteria. Both strands replicate, and the last stage of the transfer process is the reconstitution of a closed, circular duplex structure in the recipient. Many other details of this remarkable process of plasmid transfer are known, and no doubt still more will be discovered. Current evidence indicates that the process is largely controlled by plasmid genes, and that bacterial genes play little or no part (Achtman, 1973).

The complex events involving transfer and incorporation of bacterial chromosome segments, due to the activities of inserted plasmids, will not be discussed here. In general the mechanism appears to be similar to that of transfer of the F plasmid (Curtiss, 1969; Reeve & Willetts, 1974).

Recombination and mapping

Plasmid DNA undergoes a variety of recombinational processes, that is, different DNA segments associate together by specific pairing and go through a series of events resulting in the formation of new, recombinant gene sequences. These events are formally analogous to certain chromosomal phenomena in eukaryotes, such as crossing-over, translocation, deletion, etc., though it is uncertain how close the analogies are. With plasmid DNA two types of recombination may be distinguished: (1) that involving the DNA of two different plasmids, and (2) that between plasmid and bacterial DNA. These will be considered in turn. In addition, certain other structural changes in plasmid DNA are known, such as the dissociation into two parts which sometimes occurs when a plasmid is transferred from one bacterial species to another, as when R factors are transferred from *S. typhimurium* to *E. coli* (Anderson & Lewis, 1965) or from *E. coli* to *Proteus mirabilis* (Rownd et al., 1975). Such dissociation has been thought to result from the pairing of certain homologous regions (denoted *IS*) at different sites around the plasmid DNA (Hu et al., 1975) and may thus be included amongst the recombinational phenomena.

Plasmid–plasmid recombination Many examples of plasmid–plasmid recombination have been reported. For example, Watanabe and Ogata (1966) found that an R factor from *Salmonella typhimurium*, bearing genes for resistance to sulfonamide, streptomycin, chloramphenicol and tetracycline, and showing low conjugal transferability, gave rise, after being introduced into a strain of *E. coli* containing an F factor, to a new type of plasmid bearing the drug-resistance genes and also showing high transferability. The new type was assumed to have arisen as a result of recombination of the drug-resistance genes in the R factor of *S. typhimurium*, with certain genes in the F factor of *E. coli* concerned with transferability.

Hashimoto and Hirota (1966) mixed together two different chloramphenicol-sensitive R factors in *E. coli*, each derived by separate mutations from a CAP-resistant form, and obtained new resistant plasmids by recombination.

It was shown that such plasmid–plasmid recombination was controlled by the genome of the host bacterium, since recombination between chloramphenicol-sensitive R plasmids occurred in rec^+ but not in rec^- bacteria (Foster & Howe, 1971). No evidence has been published indicating control of interplasmid recombination by plasmid genes.

Plasmid–bacterium recombination Insertion of plasmid DNA into the bacterial chromosome, and the reverse—excision—are processes thought to involve a recombinational mechanism involving plasmid and bacterial DNA. It has been suggested that insertion of the circular F plasmid into the circular bacterial chromosome is by a single recombinational event (see Hayes, 1968), like that suggested by Campbell (1962) for insertion of phage DNA into the bacterial chromosome.

That recombination between an F plasmid and the bacterial chromosomes can take place is proved by the occurrence of the F-prime plasmids, in which some bacterial DNA, bearing known bacterial genes (e.g. *lac*), is inserted in the F plasmid DNA. It is believed that insertion of F into the bacterial chromosome to form the Hfr bacteria involves recombination between the insertion sequences (IS) which are homologous regions in both bacterial and plasmid DNA. The *E. coli* chromosome contains at least 17 such sites, while the plasmid DNA also contains several (Sharp *et al.*, 1972; Davidson *et al.*, 1975). (These *IS* sequences were originally so-called not because of their role in the insertion of plasmid in bacterial DNA, but because of their apparent ability to be transposed about the bacterial chromosome.)

These examples are given to indicate the variety of recombinational mechanisms in which plasmid DNA may participate, under natural conditions. As is indicated later (pp. 74–6), still further possibilities exist under laboratory conditions.

Mapping plasmid DNA A number of methods are potentially available for mapping genes on plasmid DNA. Methods based on the relative frequency of recombination between different genes, similar to those used in classical genetics, have been used to only a small extent, due to problems arising from the incompatibility between two similar plasmids placed within a single bacterium. For map construction, the method of choice is that using complementation between strains containing overlapping deletions (Willetts, 1975). This makes it possible to establish the sequences of genes but not the relative distances between them. The method is supplemented by electron microscope heteroduplex studies (Davidson *et al.*, 1975) by which the physical location, in units reckoned in kilobases, of particular points on the DNA, can be established. These methods have been applied to some F and R plasmids. In Fig. 4.1, a simplified version of the R100 plasmid map is shown. It includes the group of genes concerned with transfer, the drug-resistance genes and a number of other loci, such as those controlling insertion, replication and origin of transfer. It is interesting to note the clustering of genes concerned with particular functions.

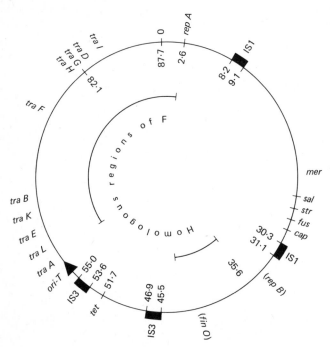

Fig. 4.1 Genetic map of plasmid R100. (Modified after Dempsey and Willets (1976), with permission.) Symbols: *traA-traL*—transfer genes; *repA, repB*—genes necessary for replication; *ori-T*—origin of transfer; IS1, IS3—insertions; *fin-O*—transfer control gene; *tet*—tetracycline; *cap*—chloramphenicol; *str*—streptomycin; *sal*—sulphonamide; *mer*—mercuric ions; *fus*—fusidic acid. Numerals inside outer circle show map distances in kilobase units. The regions of F homologous to R100 are shown by inner circle. Position of *ori-T* in R100 is not known; position given is derived from analogy with F.

The maps of the R and F plasmids so far constructed show considerable similarity, at least in regard to the sections containing the transfer genes, which seem to be closely homologous. This is indicated in Fig. 4.1 by the inner circle denoted 'homologous regions of F'.

Artificial plasmids and their use in genetic engineering

Recent developments in the use of DNA-cleaving enzymes have given bacterial plasmids a new significance in biological science. Many bacterial strains have the ability to degrade foreign DNA when that DNA is introduced into the cell. The foreign DNA may be that of a phage, introduced by the normal infective processes, or of another bacterial strain introduced by conjugation, transformation or transduction (Boyer, 1971). This cleavage is known as restriction and will occur if the incoming DNA has not been modified by passage through a strain with the same restriction/modification system as the new recipient. The cleaving enzymes may be produced as a result of activity of genes in the bacterial chromosome, in a plasmid, or in a phage. One enzyme which has been much used, denoted REcoR1, is produced by some R plasmids of *E. coli*.

Purification and analysis of the sites of action of the restriction enzymes has shown that many of them produce staggered breaks at specific sites in the DNA duplex (Bigger *et al.*, 1973; Hedgepeth *et al.*, 1972) (Fig. 4.2). This results in the generation of linear fragments with mutually cohesive ends. DNA from any source will contain, by chance, sites specific for the action of these enzymes. The fragments generated *in vitro* by the same restriction

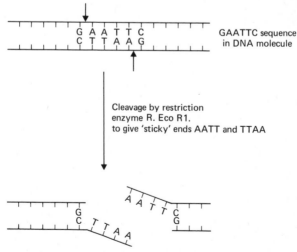

Fig. 4.2 Cleavage of DNA by restriction enzyme REcoR1.

endonuclease can be loosely joined by virtue of their cohesive ends and then covalently linked by the addition of polynucleotide ligase.

This technique opens up vast possibilities for the artificial rearrangement of genetic material. Much effort has recently been put into the development of plasmid 'vectors' for use in cloning specific segments of genetic material from various sources. The ideal vector would have only one target site for a given restriction endonuclease, enabling the covalently closed circle (CCC) of the plasmid to be opened up and the piece of DNA, which is to be cloned, to be inserted. The two DNA species are then covalently linked by means of

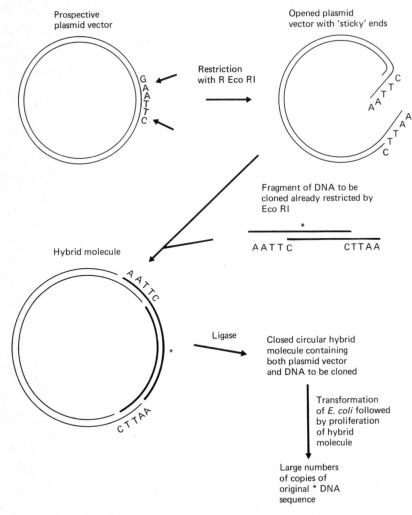

Fig. 4.3 Cloning of DNAs in a plasmid vector.

a ligase. The plasmid vector, carrying the newly inserted DNA fragment, is introduced into a suitable bacterium, usually *E. coli*, which may be grown in vast quantities and lysed, and the plasmid is then isolated by the usual simple methods for CCC DNA. The sequence of steps is illustrated in Fig. 4.3.

The isolated DNA may be used for various studies, such as structural analysis by hybridization or electron microscopy (Glover *et al.*, 1975), or functional analysis using *in vitro* transcription/translation systems. These studies may be performed on the cloned DNA either in association with the vector, or the DNA may be excised by use of the same endonuclease with which it was inserted.

A number of plasmids have been used as vectors and more are being engineered *in vitro*. Among the most useful of the naturally occurring vectors is the colicinogenic plasmid ColE1, which has a single EcoR1 target site and can amplify itself and any covalently linked DNA to 1000 copies per *E. coli* cell, if chloramphenicol is added to the culture. This enables vast quantities of cloned DNA to be produced (Hershfield *et al.*, 1974). Also many derivatives of ColE1 are being developed to make possible the use of this vector with other restriction enzyme systems (now about thirty) (Hershfield *et al.*, 1976).

While these manipulations may have no direct relevance to our understanding of naturally occurring plasmids as extranuclear genetic elements, the use of restriction endonuclease techniques will probably lead to a marked increase in our knowledge of the organization and function, not only of plasmids, but also of mitochondria and chloroplasts.

Conclusion

Plasmids may be considered as accessory genetic systems of bacteria. These systems are certainly not indispensible, since bacteria can survive without plasmids. However, the potentialities of bacteria are markedly expanded when they contain certain types of plasmid. This is most obvious, of course, in regard to the 'mating' behaviour of bacteria, if one can call it that. Put in another way, some plasmids give their bacterial hosts the 'service' of assisting them to exchange genetic material with other bacteria. Another way of looking at plasmids, however, is to consider them as independent, or partially dependent, organisms or parasites, which use bacteria as convenient ecological niches and which can move around with remarkable facility from one host to another.

The interplay of the two genetic systems, of bacteria and plasmids, is extraordinarily intimate. Indeed, when plasmid DNA is inserted in the bacterial chromosome, the two systems fuse into one. While the bacterial system is completely self-sufficient, that of the plasmid is very incomplete. It does not contain the elements necessary for constructing a protein-synthesizing apparatus, comprising ribosomal RNA and protein, t-RNAs and so forth; and in many fundamental respects, such as control of DNA replication, the plasmid is very dependent on the bacterial genetic system.

It is difficult to specify the essential minimal content of a plasmid. As stated earlier, the original classification into F, R, Col types is unsatisfactory, since the properties characteristic of each type can be lost, gained or recombined. Compatibility or ability to grow in a bacterium containing another type of plasmid may be a more convenient basis for classification.

Perhaps plasmids should be considered as a mere flotsam and jetsam of DNA floating about in the microbial world, able to become attached to, or separated from, numerous other pieces of DNA, and having only the one indispensible capacity of retaining their individuality by determining the structure of daughter DNA strands. Nevertheless, recent research has indicated some of the extraordinarily intricate details of plasmid behaviour, demonstrating their great value as models for comparison with the more complex cell organelles of eukaryotes.

5 Endosymbionts and Viruses as Agents of Extranuclear Heredity

Introduction

Some types of cytoplasmic heredity are caused by the presence of endosymbionts in the cytoplasm. We use the word endosymbiont to mean one organism living within the cells of another, and do not imply that this association necessarily confers any advantage or disadvantage on either partner, though it may do so. Symbiosis, according to one definition is 'a relatively intimate association of two or more organisms' (Ball, 1968). Here we will accept as endosymbionts intracellular inclusions showing re- semblances to representatives of known groups of microorganisms such as bacteria, algae, fungi or protozoa, many of which of course also lead a free, non-symbiotic and non-parasitic life. Some of these microorganisms contain viruses, and apart from that there may be viruses free in the cells of multicellular organisms. According to one point of view, viruses should not be considered as organisms because of their very restricted metabolic capabilities, and can therefore hardly be included within the category of endosymbionts. Nevertheless in a number of the examples of extranuclear heredity to be discussed, viruses or virus-like particles play an essential part and it is therefore convenient to include them here.

This chapter is particularly concerned with the genetic aspects of endosymbionts, and the interrelations between the genetic systems of symbionts and hosts. A case can be made out for the proposition that endosymbionts contain their own independent genetic system, which determines all the major features of the symbiont, and which operates independently of the host's genetic system. If this is so, endosymbionts are clearly in a class distinct from that of organelles such as mitochondria and chloroplasts which are so largely dependent on the nuclear genetic system of the host cell. However, such a radical distinction between organelles and symbionts has not yet been conclusively established. Some of the details given in this chapter should clarify this situation to some extent.

There are immense numbers of symbiotic organisms, derived from the whole range of animals, plants and microorganisms. The protozoa alone contain hundreds, even thousands, of examples (Kirby, 1941; Ball, 1968). Relatively very few will be referred to here. We will restrict ourselves to those which seem to bear on extranuclear heredity, that is, which affect some observable trait of the host organism, and which are transmitted from one host generation to another via the cytoplasm, or at least by some means other than the chromosomes of the host.

The kappa particles of the ciliate protozoan *Paramecium aurelia* may be taken as the prototype of a genetically significant endosymbiont. They will therefore be considered first.

Endosymbionts of *Paramecium* and other protozoa

Introduction The 'killer' paramecia were discovered by Sonneborn in 1938, when he observed that certain individuals of *P. aurelia* liberated into the water an agent capable of killing other, sensitive, paramecia. The ability to act as a killer was found to be inherited through the cytoplasm, and the cytoplasmic factor concerned was denoted kappa. As a result of these findings, arguments arose as to the nature of kappa, whether it was a cytoplasmic gene or 'plasmagene', or, on the other hand, something more like a virus or symbiont. The controversy has now been settled in favour of the symbiont hypothesis, but this has not led to a cessation of interest in the subject. Recent work has been concentrated mainly on the nature of the kappa symbionts and on the interactions between the genetic systems of symbionts and host.

Only a portion of the available data will be presented here. More extensive accounts will be found in various reviews (Sonneborn, 1959; Beale *et al.*, 1969; Preer *et al.*, 1974).

Kappa particles—general description Kappa particles (recently designated as a bacterial species and named *Caedobacter taeniospiralis* (Preer *et al.*, 1974)) are found in some naturally occurring stocks of *Paramecium aurelia*. Out of the fourteen identified 'species' of *P. aurelia* (formerly denoted varieties or syngens), kappa is found in species 2 and 4 (or *P. biaurelia* and *P. tetraurelia*, as recently renamed (Sonneborn, 1975)). Some examples are illustrated in Figs. 5.1a and b. They are usually 1–5 μm in length, bounded by a double membrane, and are Gram negative and Feulgen positive. Their DNA is not concentrated in a central 'nuclear' region, as in *E. coli* and some other bacteria, but dispersed throughout most of the particles.

Two different types of kappa denoted 'bright' and 'non-bright' can often be seen in the same paramecium. The percentage of brights is usually quite low, but rises to 35% under certain conditions. Brights contain one, or occasionally two, curious refractile bodies, which are seen by electron microscopy to consist of coiled ribbons of proteins (see Figs. 5.1a, b). The presence of brights is a diagnostic feature of kappa particles, distinguishing them from other endosymbionts of paramecium. The bright kappa particles confer on the paramecium the ability to act as a killer.

The growth rate of kappa varies according to the conditions (temperature, food concentration, etc.) and is not necessarily the same as that of the host paramecium. Hence the number of kappas per paramecium may increase or decrease, from many hundreds to a few or none. Growth rates also vary according to the strain of kappa. Under certain conditions, the growth rate

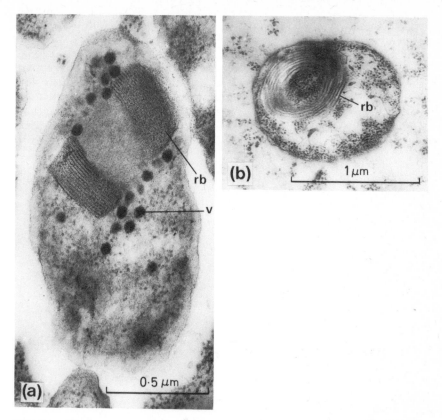

Fig. 5.1a Electron micrograph of a kappa particle in a cell of *Paramecium aurelia*, stock 562. This shows the refractile body (rb) and virons (v) found in up to 35% of kappa particles. (With permission from J. R. Preer *et al.* (1972), *J. Cell Sci.*, **11**, 581–600 and The Company of Biologists, Ltd., Cambridge.)
Fig. 5.1b Electron micrograph of a cross section of a kappa particle. The spiral structure of the refractile body (rb) can be seen. (By kind permission of A. Jurand.)

of the paramecia exceeds that of the kappa particles, and a 'curing', i.e. a change from killer to sensitive, takes place. Not much is known about the replication of kappa, except that at least one particle must be present if more are to be formed. Non-bright kappas probably reproduce by transverse fission, while brights are thought to be non-replicating. No fusion or recombinational processes have been demonstrated.

Virus-like particles in kappa Virus-like particles can be seen on the surface of the lamellae of the refractile bodies of the bright kappa particles. Different types of kappa contain different virus-like particles. In some they appear hexagonal in outline and 50–120 μm in diameter, while in others they appear to be helical structures, about 18 μm in width and of indeterminate length. These viruses, as they will be called, contain DNA in the form of covalently closed circles, of contour length 13.75 µm (see Fig. 5.2), while the main DNA component of kappa, so far as known at present, consists of long, linear strands (Dilts, 1976). No viruses have been seen in non-bright kappas. Since non-brights are thought to give rise to brights spontaneously and especially following treatment with ultraviolet light, it is possible that the non-brights contain the viruses in a latent form, akin to that of prophages or chromosomally integrated plasmids. The viruses of kappa have never been transmitted from cell to cell by artificial infection.

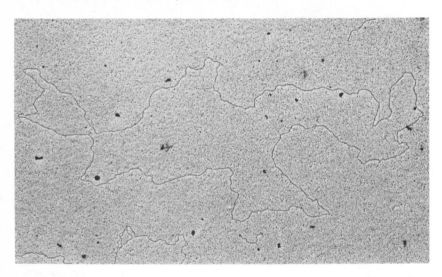

Fig. 5.2 Electron micrograph of a shadowed molecule of kappa viral DNA. The molecule is circular, covalently closed and has a contour length of 13.17 µm. (With permission from J. Dilts (1976), *Genet. Res.*, **27**, 161–170 and Cambridge University Press, Cambridge.

Effects of kappa on paramecia The bright forms of kappa particles, when present in a paramecium, make that paramecium a killer, whose lethal activities are shown by the effect on sensitive paramecia, i.e. those lacking kappa particles. The symptoms prior to death vary according to the strain of kappa, some strains causing 'humping', others spinning, others vacuolization, paralysis and so on. Death follows from the uptake by sensitive paramecia of one or more bright kappa particles, or their derivatives or components. The mechanism is not fully understood but it is possible that the viruses are the lethal agents. After death of the sensitive animals, no further kappa particles or viruses are produced. Thus the process is quite different from the lysis of bacteria by phages, whereby a new crop of phages emerges from the lysed cell. The killing process in paramecium resembles more closely that of the colicin plasmids (see p. 67), than that of bacteriophages, although, as pointed out by Preer *et al.* (1974), colicins emanating from one bacterium are toxic to other bacteria, whereas the lethal agents produced by kappa, kill paramecia, not other kappa particles.

The above account is a description of the effect of kappa particles on sensitive (i.e. kappa-less) paramecia. What is remarkable is that the presence of many hundreds of kappa particles inside a paramecium has no adverse effect on the host. Growth rate and ability to undergo the sexual processes of conjugation or autogamy are unimpaired, and the only known effect on the host paramecium itself is to render it immune to the killing effect of external kappa particles like those inside (Sonneborn, 1959).

To illustrate the bacterial qualities of kappa particles, some characteristics may be mentioned, as summarized by Preer *et al.* (1974). Kappa particles contain, apart from DNA and RNA, various enzymes, of which those of the pentose–phosphate shunt are of particular interest because they do not occur in paramecia themselves. Kappa particles have been shown to respire. They contain cytochromes a_1, a_2, b_1 and o, which are different from the cytochromes of paramecia and similar to those of certain bacteria. Kappa particles are sensitive to the effect of certain antibiotics (Sonneborn, 1959), though such tests are complicated by the effect of antibiotics on the host paramecia and food bacteria, as well as on the symbionts.

What has been described should be sufficient to establish the bacterial affinities of kappa, though of course it is not exactly like any known free-living bacterium. In fact, the ability to grow outside paramecia in an artificial medium is one important property which kappa lacks. However, certain other of the paramecium symbionts (lambda and mu) have been successfully cultivated in nutrient media, according to van Wagtendonk *et al.* (1963) and Williams (1971). Moreover, kappa can certainly be transferred from one paramecium to another by infection through the external medium. Thus Koch's postulates are at least partially satisfied.

Other endosymbionts in *Paramecium* and in other protozoa Apart from kappa, many other endosymbionts are now known in different species of the *Paramecium aurelia* group, in other species of *Paramecium* (*P. caudatum, P. bursaria*), in other ciliate genera (*Euplotes, Tetrahymena*), and in non-ciliate protozoa. These symbionts constitute a varied group of bacterium-like organisms. In *Paramecium* they range in size from the minute gamma particles (0.7 μm long) to the relatively large lambda (3 μm) (see Fig. 5.3) and sigma (up to 15 μm) particles. Most occur in the cytoplasm, but some are found in the macronucleus, e.g. the spiral alpha particles in *P. aurelia* and another type described long ago in *P. caudatum* (Petschenko, 1911). Recently a symbiont (denoted omega) has been reported in the micronucleus of *P. caudatum* (Ossipov, 1976). These may be included, somewhat illogically, in the category of 'extranuclear' hereditary factors, since they are separate from the chromosomal material in the nuclei, and are transmitted quite independently of the nuclei during the sexual phases of the paramecia.

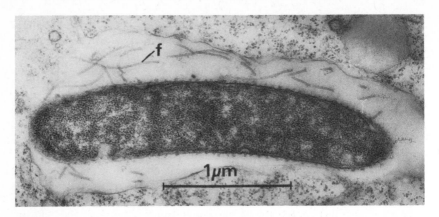

Fig. 5.3 Electron micrograph of a lambda particle in a *Paramecium* cell. The long strands (f) outside the particle may be flagella. (By kind permission of A. Jurand.)

In some *Euplotes* species, cytoplasmic endosymbionts seem to be essential for the life of the ciliate, because when these symbionts are eliminated by penicillin treatment, the ciliate dies unless reinfection is brought about within a short time (Fauré-Fremiet, 1952; Heckmann, 1975). It has been suggested that all fresh-water *Euplotes* species may contain such symbionts.

It is not known what proportion of ciliates in natural populations contain endosymbionts. The number is certainly appreciable, yet many ciliates live without symbionts. For the most part they may be considered as optional constituents of the ciliate cell, and in a few—obligatory constituents.

Turning to non-ciliate protozoa, it is worth mentioning that several amoebae are known to contain symbionts (Hawkins & Wolstenholme, 1967) and one form, *Pelomyxa palustris*, is particularly interesting in that it lacks mitochondria but contains bacterium-like symbionts (Whatley, 1976). It is conceivable, though not proved, that these symbionts may be fulfilling some of the roles normally played by mitochondria. Examples of symbionts in flagellates, e.g. in *Crithidia oncopelti*, have also been described (Spencer & Cross, 1975).

Apart from the bacterial symbionts described above, some protozoa, such as *Paramecium bursaria*, contain algal symbionts. Although most, if not all, wild individuals of this ciliate contain large numbers of zoochlorellae, it is possible to prepare *P. bursaria* lacking algae by antibiotic treatment or prolonged growth in darkness. The bleached paramecia survive loss of algae provided there is an adequate supply of bacterial nourishment in the medium. Reinfection of alga-less *P. bursaria* with chorellae is readily achieved. Some degree of mutual adaptation of ciliate and alga has been demonstrated, as judged by the varied success of different combinations of strains of both members of the partnership. Both ciliate and alga benefit from the association: the ciliate because it can grow well in water lacking bacteria or other nutrients if chlorellae are present in the cytoplasm; the alga because it can better survive growth under poor light conditions when inside a paramecium than when free in the external medium. However, the genetic basis of these interactions has not been analysed (Karakashian, 1975).

Genetic interrelations between bacterial symbionts and paramecia Since both symbionts and paramecia contain DNA, it is interesting to ask how each of these types of DNA controls the various characters of each member of the partnership, how the two genetic systems interact, and furthermore, what is the role of the DNA in the viruses within kappa. Unfortunately these are largely unanswered questions at present.

Sonneborn (1943) made a genetic analysis of the killer paramecia by making crosses between killers and sensitives and examining the progeny. The main results are shown diagrammatically in Fig. 5.4. It may seem surprising that conjugation between killers and sensitives does not harm the sensitives, but in fact it does not, provided conjugation starts shortly after the two cultures are mixed.

These experiments gave the first proof of the cytoplasmic inheritance of killing ability, which was later associated with the microscopically visible kappa particles. The crosses shown in Fig. 5.4 also provide evidence for the presence of a dominant gene, K, in the stocks of killer paramecia. Replacement of K by its allele k leads to an irreversible loss of kappa, which is not re-established when the dominant allele is reintroduced by a further cross. Such reappearances can only occur by introduction of kappa from outside, either by abnormal cytoplasmic transfer at conjugation, by infection of the paramecium from a concentrated preparation of extracted kappa particles in the nutrient medium, or by microinjection (Sonneborn, 1959; Preer *et al.*, 1974; Gibson, 1973; Koizumi, 1974). It is also of interest to note that paramecia carrying the gene K but lacking kappa are not killers and are not immune to killing by other killer paramecia. The gene K is thus necessary but not sufficient for development of kappa.

Other symbionts in different species of *P. aurelia* have been shown to require the presence of one or more dominant genes in the host paramecia. The contribution of these genes of *Paramecium* towards the maintenance of the symbionts is however not understood. Variant forms of some symbionts are known, e.g. the different types of kappa in stock 51 of species 4 of *P. aurelia*. These would provide material for a genetic analysis of the kappa particles themselves, if a suitable technique were available (e.g. by recombination between symbionts). But no such technique exists. All that can be stated is that a number of different symbiont variants can be maintained in different paramecia all containing the same paramecium genes. Hence this symbiont variation is presumably due to variation in symbiont genes, but this is speculation at present.

By microinjection it is possible to introduce various types of symbiont into paramecia of different genotypes and even different species. The results of such injection experiments have given rather confusing results, in that some particles, following injection, can apparently be maintained in paramecia lacking the supposedly necessary dominant gene (Gibson, 1973). These results require further study, but one possible interpretation is that the dominant genes referred to are only required at certain stages of the life cycle, e.g. during conjugation or autogamy. This would not conflict with the original breeding results like those shown in Fig. 5.4.

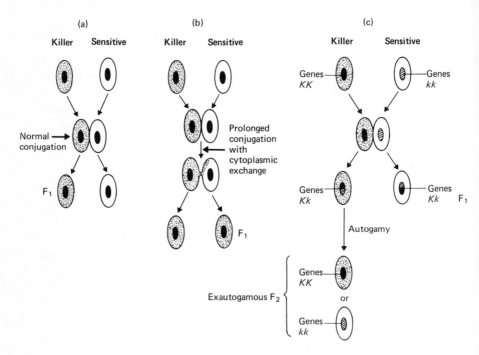

Fig. 5.4 Genetic analysis of killer paramecia. (a) Conjugation without cytoplasmic exchange. The phenotypes of the exconjugants do not change. (b) Conjugation with cytoplasmic exchange. Both exconjugants become killers. (c) Conjugation (without cytoplasmic exchange) between KK killers and kk sensitives followed by autogamy to produce homozygous F_2 cells. The killer phenotype is now seen to be due to a nuclear gene since exconjugants homozygous for the K gene remain killers but cells homozygous for the k gene become sensitive.

The general conclusions, based on the limited data at present available, is that there is very little intimate interaction between the genetic systems of symbionts and paramecia, comparable with that which occurs between plasmids and their host bacteria, or between mitochondria and their host cells. The development of a symbiont is made possible by certain general conditions provided by the host, but the specific symbiont properties are controlled by its own genetic system, not that of the host. Some properties of the paramecia are, of course, modified by the presence of symbionts within. For example, paramecia containing symbionts become immune to the lethal action of the same type of symbionts in the external medium, and paramecia containing mate-killer (mu) symbionts develop some new surface properties which are lethal to a sensitive conjugating partner. However, as mentioned previously, detailed information bearing on the activities of the symbiont's genetic system are at present not available.

Incompatibility in mosquitoes

Introduction Different wild strains of the mosquito *Culex pipiens* show various incompatibilities when hybridized. After matings involving some combinations of strains, the eggs which are inseminated either do not develop, or if they do, only to a limited degree. Depending on the strains used as parents in a cross, and on which strain is used as female and which as male parent, different results are obtained. Thus, if we formally consider two strains, A and B, the cross A♀ (female) × B♂ (male) might produce viable offspring, and the reciprocal cross B♀ × A♂ might be sterile; or inviable progeny might result from both types of cross, or again both might produce viable progeny, or finally certain crosses might lead to a reduced proportion of viable progeny. All of these situations arise from different crosses, depending on the strains used.

The basis of these incompatibilities is, in some cases at least, connected with a genetic factor which is inherited through the cytoplasm of the egg, and which influences the fertilizing ability of sperm in an insect developing out of such an egg. There is now some evidence suggesting that this cytoplasmic genetic factor may be a rickettsia-like microorganism, and this is the reason for including an account of mosquito incompatibility in a chapter on endosymbionts. The main genetic analysis has, however, been carried out without any reference to an endosymbiotic determinant (Laven, 1959, 1967).

Genetic analysis An extensive genetic analysis of the compatibility relationships between *Culex pipiens* strains from different parts of the world has been made by Laven. To illustrate the findings, the results of crosses between two strains denoted Ha (Hamburg) and Og (Oggelshausen) will be described. The cross Ha♀ × Og♂ gives viable offspring, about 87% of the eggs developing into adults. The reciprocal cross Og♀ × Ha♂ yields very few adult progeny, though a considerable proportion of the eggs undergo a limited development, to a stage corresponding to about 32 hours embryogenesis, but fail to develop further. A very few (about 0.17%) of the eggs derived from the cross Og♀ × Ha♂ develop into adults, but these are all females, show exclusively maternal characteristics and are considered to be parthenogenetic offspring. The precise stage at which failure of development occurs following a mating between Og♀ × Ha♂ is not certain. The sperm enter the eggs, 'fertilized' eggs are laid, and as already stated may start to develop, but union of male and female gamete nuclei does not take place.

The cytoplasmic inheritance of the compatibility characteristics of a strain was shown from the results of crosses of the type

$$Ha♀ \times Og♂$$
$$\times Og♂$$
$$\times Og♂ \ldots \text{etc.}$$

i.e. by a series of backcrosses involving females of hybrids and males of strain Og. This was continued for as many as 60 generations, by which stage all the nuclear genes of the Ha strain would be expected to have been replaced by Og genes, while any cytoplasmic factors which might have been present in the Ha parent would be retained. By incorporating marker genes in the experiments, Laven showed that normal Mendelian behaviour of other characters (controlled by nuclear genes) was taking place.

Males from any backcross generation up to the 60th, when used to inseminate Og females, produce inviable offspring (except for rare exceptions interpreted as due to parthenogenesis), while male or female mosquitoes from any backcross generation continue to give viable progeny when mated with mosquitoes of the Ha strain. These results show that some cytoplasmic factor in the Ha strain maintains its integrity unaltered in the presence of Og nuclear genes. Males developing out of eggs containing this factor produce sperm which is incapable of causing eggs lacking the factor to undergo normal development.

The nature of the cytoplasmic factor was not identified by Laven (1967), who rejected suggestions that it might be a virus or symbiont. The reason for this rejection was based on the large number (at least seventeen) of different incompatibility groups found, which would, it was believed, require at least seventeen different types of the supposed symbiont. Recently, however, other workers have obtained evidence that a rickettsia-like microorganism may be responsible (Yen & Barr, 1971, 1974). Such particles, named *Wolbachia pipientis*, were observed in *Culex pipiens* by Hertig (1936) and probably occur in practically all individuals of wild races of this species. In size the rickettsiae range from 0.25 μm in diameter and 1.5 μm in length to large coccoid forms over 1 μm in diameter, but few ultrastructural studies have been made. The rickettsiae occur in the follicles of the mosquito ovary and are especially abundant near the micropile.

By treating larvae with the antibiotic tetracycline, Yen and Barr eliminated the rickettsiae and then carried out hybridization experiments with mosquitoes lacking endosymbionts. It was found that two strains which had previously been incompatible when the rickettsiae were present, were now compatible. Males lacking rickettsiae were found to be compatible with all females, with or without rickettsiae, while males containing rickettsiae were unable to produce viable offspring with females lacking rickettsiae. It therefore seems to be the presence of rickettsiae in the males that is responsible for the incompatibility, while inheritance of the rickettsiae is through the females. Yen and Barr proposed that each strain of mosquitoes contains its 'own' co-adapted strain of rickettsiae. When a male contains rickettsiae of a type different from those in the female, or when the female has none, the action of the sperm is changed in such a way as to prevent the formation and development of diploid eggs. Whether the rickettsiae have any other action—favourable or unfavourable—on their insect hosts is not known; nor is anything known about the action of the genes of the mosquito on the maintenance of the rickettsiae except that Laven's backcrosses show that genes of any one strain are not required for maintenance of its 'own' cytoplasmic type.

Sex ratio in *Drosophila*

Introduction A number of *Drosophila* species are known to produce strains which yield male offspring in proportions much below the normal 50%, and sometimes none at all. This condition, known as 'sex ratio' (SR) is controlled by an agent transmitted through the egg cytoplasm, and in some species, though not all, SR is correlated with the presence of helical, wall-free prokaryotic organisms, formerly thought to be spirochaetes, but more recently identified as spiroplasmas (Williamson & Whitcombe, 1974). These spiroplasmas contain viruses, whose role in the male-killing mechanism is however not clear. One species of flies lacking the spiroplasmas altogether may, nevertheless, fail to produce male offspring, and it has been suggested that in this species viruses unconnected with spiroplasmas are involved in eliminating the male zygotes. The situation varies in different *Drosophila* species (Poulson, 1963; 1968; Oishi & Poulson, 1970).

Sex ratio in *Drosophila willistoni* and related species All flies of *D. willistoni* showing the SR condition are derived from one female collected in Jamaica. Strictly maternal inheritance of SR is shown, and repeated backcrossing of SR females to males of some normal strains, such as one denoted Barbados-3, results in permanent maintenance of the SR condition: with rare exceptions, only females are produced. However, repeated backcrossing to other, so-called 'disrupter', strains—e.g. one called Recife-3—leads to a loss of the SR condition. It is concluded that SR is due to the presence of a cytoplasmic factor, but this in its turn requires the presence of some *Drosophila* nuclear genes which are absent in the disrupter strains. The cytoplasmic factor kills the male zygotes, but usually not the female zygotes. In some genotypes and under some conditions, however, even the females are killed by the SR factor.

The SR factor of *D. willistoni* can be transmitted by infection. If haemolymph from SR females, or extracts of dying zygotes or larvae of SR strains, are injected into the abdomens of virgin females of normal strains, these females are converted into SR individuals after an incubation period of 12–14 days and are then able to beget only female progeny. SR factors can also be transferred from *D. willistoni* into other *Drosophila* species, such as *D. melanogaster*, and the injected recipients then produce fewer than normal male offspring

If SR flies of *D. willistoni* are examined by phase-contrast microscopy, large numbers of spiroplasma-like microorganisms can be seen in various tissues and in the haemolymph. These microorganisms are usually 4–5 µm long and were formerly identified as belonging to the spirochaete genus *Treponema* (see Fig. 5.5). No spiroplasmas are found in flies of normal male-producing strains, and there is a regular correlation between presence of spiroplasmas and the SR condition.

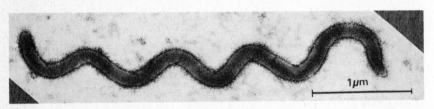

Fig. 5.5 Electron micrograph of a negatively stained spiroplasma isolated from *Drosophila willistoni*. *D. willistoni* SR flies contain large numbers of these spiroplasma-like microorganisms whereas wild-type flies do not. (In the proceedings of a conference organized by J. F. Duplan and published by Institut National de la Santé et de la Recherche Médicale, Paris. With their permission and that of Williamson and Whitcombe (1974).)

Three other species of *Drosophila*, *D. paulistorum*, *D. equinoxalis* and *D. nebulosa*, which are related to *D. willistoni*, contain strains showing similar SR phenomena, with minor variations.

The spiroplasmas within a given species of *Drosophila* may vary. For example, a strain of spiroplasmas from *D. willistoni* was selected so as to be able to grow in the disrupter strain Recife-3, which as mentioned above is normally incapable of maintaining the normal spiroplasmas from *D. willistoni*. Variation between spiroplasmas naturally occurring in different *Drosophila* species has been shown by introducing spiroplasmas from three strains—*D. willistoni*, *D. nebulosa* and *D. equinoxalis*, into the common genetic background of *D. melanogaster*. The three strains are then found to vary in temperature sensitivity and other characteristics. Maintenance of the symbiotic relationship thus requires a harmonious interplay of the spiroplasma and *Drosophila* genetic systems, though nothing is known about the genetics of spiroplasmas.

When spiroplasmas from *D. willistoni*, *D. nebulosa* and *D. equinoxalis* are mixed together, either *in vitro* or *in vivo*, they immediately form clumps and eventually disappear. The mixture made *in vivo* eliminates the SR condition. Evidence has been obtained indicating that these effects may be due to viruses present in the spiroplasmas, and it has been suggested that each strain of spiroplasma contains specifically adapted viruses which lyse other strains. When extracts were prepared from *D. nebulosa*, centrifuged to eliminate the spiroplasmas and the supernatant added to spiroplasmas from *D. willistoni*, the latter were lysed (Oishi & Poulson, 1970). The extracts contained spherical DNA-containing particles 50–60 nm in diameter, and these were shown to have entered and multiplied in the spiroplasmas of *D. willistoni*. However, proof that these viruses are concerned with the killing of male zygotes of *Drosophila* has not been obtained.

Other evidence indicates that the killing agent is separable from the intact spiroplasmas. In *D. robusta* transmission of spiroplasmas to the progeny is unusual in that it occurs only in early broods. Later broods contain few or no spiroplasmas, yet development of male zygotes is nevertheless prevented from taking place in these later broods. Hence failure of the male zygote to develop must have been due to some factor other than the intact spiroplasma (Williamson, 1966). Moreover, in other experiments (Oishi, 1971) when a virus preparation derived from spiroplasmas of *D. nebulosa* was injected into spiroplasma-containing flies of *D. melanogaster*, the recipients lost their spiroplasmas, but retained their SR male-killing properties. Hence, here also the SR effect was thought to be due to something other than the intact spiroplasmas.

Sex ratio in *Drosophila bifasciata* In this *Drosophila* species an appreciable proportion of the wild individuals from populations in Italy and Japan show the SR condition, whereby affected females mated with any male of the same species produce less than 5% male progeny (if reared at 21°C). In some respects the situation in *D. bifasciata* differs radically from that in *D. willistoni* however. In *D. bifasciata* the condition is temperature-sensitive and no spiroplasmas are present in the SR flies. According to Leventhal (1968), an intracellular virus-like agent present in *D. bifasciata* is transferable to other species and produces heritable reductions in the proportions of male progeny.

Conclusion The SR phenomenon in *Drosophila*, though clearly associated with spiroplasma-like symbionts in some species, may possibly be caused by viruses, either associated with spiroplasmas, as in *D. willistoni*, or not, as in *D. bifasciata*.

Sigma virus and carbon dioxide sensitivity in *Drosophila melanogaster*

Introduction In 1937 it was reported that a certain French strain of *Drosophila melanogaster* was unusual in being sensitive to carbon dioxide (L'Héritier & Teissier, 1937). After exposure to CO_2, the flies became paralysed and died, whilst normal flies treated in the same way were merely anaesthetized and recovered completely when returned to normal air. Many other strains of *Drosophila*, both from France and elsewhere, were also found to be CO_2-sensitive (L'Héritier, 1970).

Genetic analysis showed that the CO_2-sensitive condition, though hereditary, was non-Mendelian, and evidence soon came to light (L'Héritier & Hugon de Scoeux, 1947) implicating a virus-like agent, which was denoted sigma. However, sigma did not possess all the characteristics then considered to be typical of viruses. For example, (1) sigma has no obvious pathological effect on flies living under normal atmospheric conditions, (2) sigma is not naturally contagious, though it can be transmitted by artificial injection, and (3) it is transmitted in, or associated with, the germ cells of *Drosophila*. It was therefore put in a borderline category of agents, transmissible both by heredity and by infection and having some properties of both genes and viruses. These criteria seem less distinct to us today, now that, by special techniques, even nuclear genes can be transported from eukaryotic chromosomes into plasmids and thence into bacteria. In any case, recent work has shown that sigma is undoubtedly a virus, very similar to a known disease-causing agent—vesicular stomatis virus (VSV)—which causes a mild infection in horses and cattle and is thought to be transmitted by horseflies and mosquitoes (Printz, 1973).

However, sigma is also an extranuclear genetic component of *Drosophila* and will be considered here from that point of view.

Genetics of carbon dioxide sensitivity   The inheritance of CO_2 sensitivity in *D. melanogaster* is as follows. Females of the originally isolated strain (denoted 'stabilized' CO_2-sensitives) transmit the condition to nearly all their offspring, regardless of whether the male parent is sensitive or resistant, and CO_2 sensitivity is maintained after a series of backcrosses of CO_2-sensitive females to normal males. The condition is also inherited to some extent from male parents, in that CO_2-sensitive males, mated to normal females, produce a small proportion of CO_2-sensitive progeny; and taking CO_2-sensitive females from the latter, a few transmit the condition further. However, males which have derived CO_2 sensitivity from their fathers do not transmit it to any offspring. This is the situation in *D. melanogaster*. In some other *Drosophila* species, males may transmit CO_2 sensitivity regularly (Williamson, 1961).

No linkage has been detected between CO_2 sensitivity and characters controlled by genes on any of the four chromosomes of *D. melanogaster*. Thus, cytoplasmic inheritance is indicated, with regular transmission through the egg, and sporadic transmission through the sperm. However, many variants of these patterns of inheritance have been described.

Transmission of CO_2 sensitivity by infection was shown to occur by L'Héritier and Hugon de Scoeux (1947) by inoculating normal flies with extracts from CO_2-sensitive flies. The recipients developed the CO_2-sensitive condition and in the case of females transmitted it to their progeny. The details of the subsequent inheritance of this CO_2 sensitivity acquired by infection are complex, and will not be gone into here.

Variants are known both of sigma and of *Drosophila*, affecting the ability of sigma to grow in *Drosophila*. The sigma variants are listed in Table 5.1.

Table 5.1 Variants of sigma virus in *Drosophila* (after Seecof, 1968)

Symbol	Description
ts	temperature-sensitive
P^+	survives in presence of *Drosophila* gene *ref-2*
rho	produces CO_2 sensitivity in only small fraction of flies, and after long time
ultra-rho	does not produce CO_2 sensitivity, but makes flies immune to infection by normal sigma virus
g^+	in female flies, invades germ line by germinal passage
v^-	fails completely to enter sperm
'strong'	a single infectious unit quickly makes flies CO_2-sensitive.

Their exact genetic basis is unclear, since no technique exists at present for their analysis.

Some attempts have been made to demonstrate recombination between different sigma characters. After flies were infected with mixtures of two types of sigma, differing in a number of genetic markers, usually only the two parental types were recovered subsequently, but some indications of recombination were obtained in an experiment where females were inoculated with one virus type and then crossed with males transmitting another type of sigma. The data were, however, too few for definite conclusions to be drawn.

Several *Drosophila* mutants are known to slow down or inhibit sigma multiplication. One is denoted 'refractory' (*ref-2*), and is due to a semi-dominant gene located on chromosome II. In the homozygotes (*ref/ref*), multiplication of sigma (except mutant $P+$) is stopped, though the virus is not destroyed, since reintroduction of the wild type allele of the gene *ref-2*, by a subsequent cross, allows the virus to resume growth. Heterozygotes (*ref-2/ +*), when infected by sigma, permit growth of sigma, but at a slower rate and with a lower final yield of infective units than the homozygous wild type. As mentioned above, virus mutant $P+$ grows even in flies homozygous for the gene *ref-2*. These interactions between sigma and *Drosophila* mutants have been confirmed by experiments with cultured cells of *Drosophila*. It has been suggested that the allele *ref-2* of *Drosophila* might act on the synthesis of viral RNA (Richard-Molard, 1975). Though this suggestion is speculative at present, these variants make possible an investigation of such interactions, and ultimately it may be possible to determine in what respects growth of sigma is made possible by activities of the *Drosophila* genes.

The viral nature of sigma By studying the material necessary for transmission of CO_2 sensitivity by infection, it has been deduced that sigma is a virus. For example, from X-ray inactivation and filtration studies, the infecting agent was estimated to be a particle of diameter 180 nm (Seecoff, 1968). The growth curves of the infectious particles were plotted by preparing extracts at known times, making a series of dilutions of the extracts and determining the minimum amount of material required to establish new infections. A single infectious unit is all that is necessary, under favourable conditions, to transmit CO_2 sensitivity to a new fly. Many studies have been made on the course of events following infection, using the above-mentioned titration method.

L'Héritier (1958) formulated the distinction between 'stabilized' and 'non-stabilized' CO_2-sensitive flies. In the former, inheritance through females is highly efficient and through the males only sporadic; in the latter, females transmit the condition to a proportion of their offspring, males to none. The material basis for these differences has not been established: presumably it is connected with variations in the ability of the virus to become established in the germ lines. It is known that males infected with virus from their father's sperm contain infectious material in the somatic tissue of the testis, but not in the sperm itself. Where infection is through the males, the virus is thought to lie within the sperm, not merely in the seminal fluid. It was at one time suggested that the difference between the stabilized and non-stabilized states was analogous to that between prophage and vegetative phages, and that stabilization was analogous to lysogenization (L'Héritier, 1955), but no evidence for the integration of sigma virus within the *Drosophila* chromosomes has been put forward.

While these studies strongly supported the hypothesis that sigma is a virus, the matter was conclusively settled by obtaining visual evidence by electron microscopy of viruses in CO_2-sensitive flies, and the absence of such viruses in normal flies (Berkaloff *et al.* 1965; Printz, 1973). Sigma is then seen to be a bullet-shaped particle, of length 140–180 nm and diameter 70 nm (see Figs. 5.6, 5.7). It is very similar to VSV (Fig. 5.8), though the latter is slightly longer than sigma. Both have the same shape, characteristic cross-striations and surface projections. Both grow by budding from cellular membranes of the host. VSV is known to contain a single-stranded RNA and three major proteins; as for sigma, direct proof that it contains RNA (rather than DNA) has not been established, due to difficulties in purifying the material, but it is assumed that sigma also contains RNA.

VSV (the Indiana strain) has been inoculated into *Drosophila* and in that habitat follows a similar growth curve to that of sigma, but VSV does not immediately cause the fly to become CO_2-sensitive. No virus can be recovered from the progeny of VSV-inoculated flies. However, after a period of cultivation in *Drosophila*, a VSV variant was obtained which did confer CO_2 sensitivity. Sigma also differs from VSV in that, while VSV can be cultivated in vertebrate cells, sigma cannot. Thus, there are minor differences between VSV and sigma, but in general they are very similar, and both are classified in the group of rhabdoviruses.

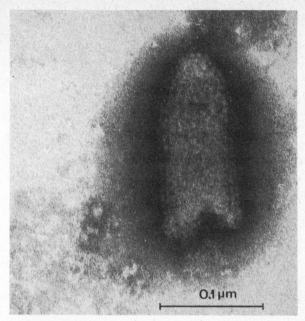

Fig. 5.6 Electron micrograph of a negatively stained sigma particle. (By kind permission of D. Tenninges.)

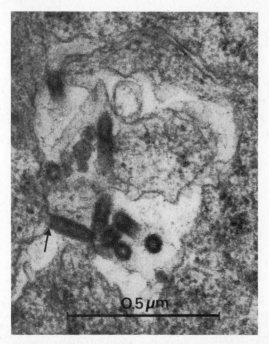

Fig. 5.7 Electron micrograph of a section through *Drosophila* spermatids. The arrow indicates a sigma particle in the process of budding on the cytoplasmic membrane of the host cell. (With permission from D. Tenninges (1972), *Annales de l'Institut Pasteur*, **122**, 541–67 and Service des Annales de l'Institut Pasteur.)

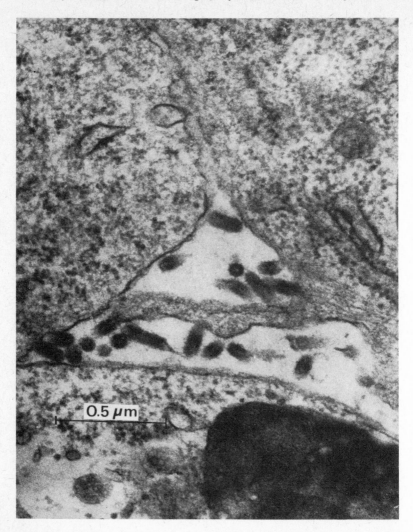

Fig. 5.8 Electron micrograph of a section through chick embryo fibroblasts showing vescicular stomatitis virus (VSV.) budding from host cell membranes. (By kind permission of D. Tenninges.)

Effects of sigma on Drosophila The physiological effects of CO_2 on sigma-containing flies are not clearly understood, though the thoracic ganglia are thought to be the regions mainly involved. Sigma usually has no marked effect on its host, though under certain conditions (apart from presence of excess CO_2) there may be some disadvantages. Very high concentrations of sigma may slow down the development of larvae and pupae and result in the laying of fewer eggs (L'Héritier, 1970). As stated earlier, a significant proportion of the flies in wild populations of *D. melanogaster* contain sigma. Since its inheritance is by no means 100% certain, and there is apparently no natural infectious transmission or growth in vertebrate cells, possession of sigma may, one suspects, confer some compensating advantage to the host flies.

Conclusion The finding that sigma, once considered a cytoplasmic gene in *Drosophila*, resembles a known pathogenic virus of vertebrates (VSV), is rather surprising. It should, however, be remembered that VSV is also transmitted through insects. It may well be that there are many examples of non-contagious viruses which are transmitted through the germ cells of insects and which have pathological effects on vertebrates. Sigma was only discovered because of the intensive study of *Drosophila* by geneticists, who need to anaesthetize their flies during their experiments.

Double-stranded RNA and killer strains of *Saccharomyces* and *Ustilago*

Saccharomyces cerevisiae Some strains of yeast have been found to secrete a protein-containing substance which kills sensitive yeast cells. These 'killer' yeasts contain a cytoplasmic factor which is absent in ordinary sensitive cells. Apart from killers and sensitives, a third class of yeast cells, denoted 'neutrals', is known. These are neither killed by the agent secreted by killers, nor do they themselves kill sensitives. 'Neutrals' also contain a cytoplasmic factor. Table 5.2 summarizes the genetic evidence for these statements.

Table 5.2 Inheritance of killer yeasts (after Somers & Bevan, 1969)

Parent strains	Diploid hybrids	Tetrads of hybrids
Killer × Sensitive	Killer	4:0 Killer:Sensitive
Neutral × Sensitive	Neutral	4:0 Neutral:Sensitive
Killer × Neutral	Killer	Various ratios from 4:0 to 0:4 of killers and neutrals. No sensitives

Maintenance of both killer and neutral cytoplasmic factors was shown to require the presence of a dominant nuclear gene denoted M(later $+^{mak-1}$); in later work, further nuclear genes were discovered, e.g. *kex 1* and *kex 2* which convert killers into neutrals (Wickner & Leibowitz, 1976).

All killer and neutral strains have been found to contain two species of double-stranded RNA (a larger and a smaller), separately encapsulated in isometric virus-like particles, having a mean diameter of 39 nm (Herring &

Bevan, 1974). (Fig. 5.9.) It is believed that the smaller RNA-containing particles are necessary for toxin production: they are dependent upon the presence of both the $+^{mak-1}$ allele and the larger RNA-containing particles. The latter are inherited cytoplasmically and independently of the host genotype (Mitchell *et al.*, 1976). Some further details have been reported by Sweeney *et al.* (1976).

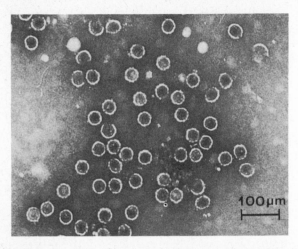

Fig. 5.9 Electron micrograph of negatively stained virus-like particles isolated from *Saccharomyces cerevisiae*. The particles have a mean diameter of 39 μm and can be isolated from both killer and neutral strains of yeast. (With permission from A. J. Herring and E. A. Bevan (1974) and Society for General Microbiology.)

Ustilago maydis This fungus, a basidiomycete causing a smut disease of maize, also produces 'killer' strains, which resemble those of yeast in that inheritance of the condition is cytoplasmic, and associated with virus-like particles containing double-stranded RNA (Fig. 5.10). The asexual, haploid, stage of the fungus can be grown on nutrient agar, producing yeast-like cells, but the sexual stages, including formation of zygotes, meiosis and production of tetrads, require cultivation of the fungus on maize seedlings, in which a mycelial type of growth develops.

Three different killer strains (denoted P1, P4, P6) have been identified, each producing a toxin which kills or inhibits growth of a sensitive strain (P2), or of the other types of killers. In addition, various 'immune' strains are known, equivalent to the 'neutral' yeasts (see above). The immune strains neither kill, nor are killed and are specifically resistant to one type of killer strain (e.g. strain P3[1] is immune to killing by strain P1 (see Table 5.3)). Genetic analysis shows that each of the killer strains, and some of the immune strains, each contain specific cytoplasmic genetic factors. In addition, a number of nuclear genes (formerly denoted s^+, later p_4^r p_4^r, p_6^r) are known and convert the sensitive strain to resistance (to one or other killer

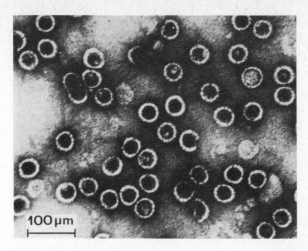

Fig. 5.10 Electron micrograph of negatively stained, virus-like particles isolated from *Ustilago maydis*. These particles can be isolated from both killer and immune strains and have a mean diameter of 41 μm. (With permission from Wood and Bozarth (1973) and The American Phytopathological Society.)

strains), even in the absence of the cytoplasmic factors for immunity (Puhalla, 1968; Koltin & Day, 1976).

Proof of the cytoplasmic inheritance of the killer character in *Ustilago* is shown by data summarized in Table 5.3. In addition, confirmation was obtained from a study of heterokaryons between killer and sensitive strains, each marked with known nuclear genes. Some of the homokaryons extracted from such heterokaryons combined the killer character from one strain with a nuclear gene from the other.

Some further details may be mentioned. Although the various killer and immune strains are stable on agar media, there is some instability when the cultures are grown in maize seedlings. Killers may give rise to immunes and sensitives, and immune strains may give rise to sensitives. Sensitives, however, are completely stable. Crosses between different killer types have been made, and show an 'exclusion' of one or other, or sometimes both, cytoplasmic killer factors. For example, P4 killers, when crossed with P6 or P1 killers, produce mainly P4 killer progeny, while crosses between P1 and P6 killers result in 'exclusion' of both killer types, and yield only non-killer progeny, which however are still immune to killing by P6 killers. No evidence for genetic recombination between the different killer factors, to produce killer types of a new specificity, was found.

Each of the killer and immune strains (except immunes due to the nuclear genes mentioned above) contain spherical virus-like particles, c. 41 nm in diameter, which are absent in the sensitive strains (see Fig. 5.10). Some of these particles have been shown to contain double-stranded RNA (Wood & Bozarth, 1973), as do all other fungal viruses so far examined (Lemke &

Table 5.3 Inheritance of killer characters in *Ustilago maydis* (after Koltin & Day, 1976)

Haploid parent strains	Diploid hybrids	Haploid progeny			Tetrads
		Killers	Immune	Sensitive	
Killer (P1) × Sensitive (P2)	All killers	94%	1%	5%	all 4 spores killers (most in this class) or all 4 spores sensitives or all 4 spores immune
Immune (P3[1]) × Sensitive (P2)	All immune	0%	68%	32%	all 4 spores immune or all 4 spores sensitive

Nash, 1975). The RNA in different killer and immune strains shows some variation. Electrophoresis on 5% acrilamide gels produces a number of RNA components, of molecular weight varying between 0.46×10^6 and 4.7×10^6. Some of these components (e.g. two species of molecular weights 2.6×10^6 and 2.9×10^6, respectively) are common to all three killer types, but other components are specific for one or other type. Immune strains derived from a given killer strain (by crosses of killers by sensitives) appear to retain some RNA components, lose others and even gain others. Details of how these various RNA components are embodied in individual virus-like particles are not yet known.

Conclusion

In this chapter a miscellaneous assortment of endosymbionts and viruses or virus-like particles, all acting as agents of extranuclear inheritance in the host organisms, has been described. These agents range from algae, bacteria (some containing viruses) and rickettsiae to various viruses not incorporated in any prokaryotic microbe. In considering this sample, one gains the impression that what has been described is but a minute fraction of the total numbers which exist. Kappa, when first discovered, seemed to be an example of a very unusual phenomenon occurring in a few strains of *Paramecium aurelia*. Now it appears that almost every ciliate species and many other protozoa consist of populations containing some symbiont-bearing individuals, and kappa-like symbionts occur also in organisms other than protozoa. Again, the sigma virus of *Drosophila*, when first discovered by its effect on CO_2-sensitive properties of the host flies, seemed a rarity; yet many strains of *D. melanogaster* are now known to contain sigma, which is realized to be but one example of the class of widely occurring rhabdoviruses. Finally, the double-stranded RNA viruses in the killer yeasts and in *Ustilago*, are representatives of a large group of similar viruses occurring in many fungi (Lemke, 1976). Hence we have observed only the small tip of a very large iceberg.

It has not been possible to include in this account any mention of the bacteriophages, or indeed of viruses in general, except a few which have special effects on the host's characteristics and are transmitted through the host's germ cells.

In view of the smallness of our sample in relation to the total numbers of symbionts and viruses existing, it seems unwise to attempt any major generalizations at this stage. However, it may be worth pointing out that so far there is no evidence for the integration of the genetic material of a symbiont into the DNA of a host cell, like that occurring with some bacteriophages and plasmids. On the other hand, there is equally no evidence showing that such integration could not occur, and in fact very few of the examples of endosymbionts studied are technically suitable for study of this problem.

In Chapter 7, we will discuss the suggestions that all organelles may have evolved from some symbiont. From the facts described here, it is evident that the possible raw material for such evolutionary processes is very diverse; one should not restrict the argument to known familiar microorganisms like *E. coli*, as possible candidates to be the ancestors of mitochondria.

6 Miscellaneous Examples of Extranuclear Inheritance

Introduction

The other chapters of this book concern known material structures of cells having genetic properties. A large number of examples of extranuclear inheritance are known, however, which have so far not been shown to be connected with any particular organelle or cytoplasmic particle, though doubtless future work will eventually establish such connections. Examples of these incompletely analysed phenomena are found particularly abundantly amongst the fungi. In some cases, where inheritance through the sexual stages cannot be studied, as in the *fungi imperfecti*, one is not even certain that the variations under consideration are truly genetic, controlled by some gene-like determinant, most likely comprising DNA, rather than some developmental or 'epigenetic' change of cells In this chapter, a selection of these examples will be considered and discussed separately. More information about those which occur in fungi can be obtained from various reviews (Srb, 1965; Jinks, 1966; Esser & Kuenen, 1967; Fincham & Day, 1971).

Fungi

The 'barrage' incompatibility phenomenon in *Podospora anserina* When certain strains of this ascomycetous fungus are cultured together on an agar plate, their hyphae do not freely anastomose, but die back after briefly making contact, leaving a clear area or barrage in the regions where the two strains meet.

It is possible to hybridize two barrage-forming strains by taking conidia from one strain and spreading them over the protoperithecia of the other. Since conidia transmit little or no cytoplasm to the zygotes, the genetic differences between two strains can therefore be analysed in regard to cytoplasmic, as well as nuclear factors (Rizet, 1952; Beisson-Schecroun, 1962).

When two strains S and s, which form a barrage when they come into contact, are crossed, the perithecia formed give rise to asci containing two spores of one type and two of another. However, the products of the cross are peculiar in that while two are phenotypically like S, the other two yield a type denoted s^S, which does not form a barrage with either S or s strains. Crosses between s and the new type s^S show that the difference is cytoplasmic, since inheritance is strictly 'maternal'. The type s^S is not completely stable, and is rapidly transformed to s following contact with an s strain, due not to the passage of nuclei between the strains, but to a kind of 'infection' by some cytoplasmic component. Spontaneous change from s^S to s may also occur at a low rate. The reverse change, from s to s^S can occur without mating of strains with S, when very small fragments of s-type mycelium are prepared and allowed to regenerate (Belcour, 1975).

These results show that strains S and s differ not only in regard to a pair of nuclear allelic genes (S/s), but also in some cytoplasmic factor. The gene S, in the zygote of genotype Ss, is thought to eliminate or reduce to a low level the amount of the cytoplasmic factor derived from strain s, and it is the presence of this cytoplasmic factor which causes the incompatibility with strain S. When the zygote (Ss) segregates spores containing the allele s, the cytoplasmic factor from strain s is lacking and therefore the type s^S is produced, but can be converted to s by infection with the cytoplasmic factor from s type hyphae.

Strain s therefore contains a cytoplasmic factor which can co-exist with the nuclear gene s, but not with its allele S. The nature of the cytoplasmic factor is unknown.

Senescence in *Podospora anserina* Many fungi show ageing phenomena. After a period of asexual growth at the maximum rate, growth slows down and may stop altogether, and other characters, such as the ability to undergo sexual reproduction, may also deteriorate. In the absence of the sexual process, or other rejuvenating stimulus, a culture may die out.

In *Podospora anserina*, senescence is manifest by a slow growth rate, formation of slender, undulating hyphae, vacuolated and swollen hyphal tips, and eventually death. However, senescent cultures do not necessarily lose the ability to undergo sexual processes. If conidia from a senescent strain are used to fertilize protoperithecia of a normal strain, all the progeny are normal; but the reciprocal cross produces cultures which are senescent or have reduced longevity. Normal strains may sometimes be produced from senescent protoperithecia. These results indicate a transmission of the agents for senescence through the cytoplasm of the protoperithecia.

Senescence in *Podospora* is also 'infectious', as shown by experiments in which hyphae from senescent and normal strains, each containing nuclear marker genes, are placed in contact by micromanipulation. After anastomosis has occurred, fragments of the originally normal hyphae, after excision and culturing, are found to have reduced longevity. Reversal of senescence, or rejuvenation, can be induced by various treatments, such as fragmentation of the senescent mycelium, storage at 3°C, or storage under oil or under conditions of dessication.

It is believed that senescence in this fungus is caused by some cytoplasmic determining factors. It is also thought that the time of onset of senescence is due to cytoplasmic factors. Since such factors are transmitted through the protoperithecia, under natural conditions there must be an automatic destruction or counter-selection of the factors, since otherwise all cultures would become senescent and die out. The nature of these cytoplasmic factors is unknown. (For further details, see Marcou, 1961; Smith & Rubenstein, 1973a & b.) It has been suggested that senescence like that occurring in *Podospora* might be due not to a cytoplasmic genetic factor at all, but to the accumulations of errors in the synthesis of proteins (Orgel, 1963; Holliday, 1969). Such a hypothesis is rather difficult to accommodate with the maternal inheritance and infectivity of senescence.

Senescence in other fungi More or less similar ageing phenomena have been described in other fungi. In *Pestalozzia annulata* for example, senescence is transmissible by infection of a cytoplasmic factor (Chevaugeon & Digbeu, 1960), and in *Aspergillus glaucus* a phenomenon called 'vegetative death' has been described (Jinks, 1959). There is an irreversible cessation of growth, death of hyphal tips and formation of a brown pigment. In this fungus cytoplasmic inheritance cannot be shown by the results of reciprocal matings, but heterokaryosis can be exploited for the analysis. Heterokaryons involving a senescent colony with white spores and a vigorous colony with buff spores were made, and some of the extracted homokaryons had buff spores and showed vegetative death, due to a reassortment between the nuclear genes for spore colour and the cytoplasmic factors for vegetative death. Vegetative death could also be induced in vigorous colonies by contact with inocula from dying cultures, presumably by infection. Cultures of *A. glaucus* propogated regularly by sexual spores rarely, if ever, show vegetative death (Jinks, 1966).

In *Neurospora crassa* cultures grown continuously for long periods show a 'stop-start' behaviour: there are periodic interruptions in growth, but no permanent cessation (Bertrand *et al.*, 1968). A series of mutants, denoted *stp-A*, *stp-B*, etc., characterized by different frequencies of stopping were isolated, and cytoplasmic heredity was indicated, from studies involving heterokaryons, failure to transmit the characters through conidia used as male elements, and maternal inheritance through protoperithecia. From the finding that some *stp* mutants show defects in mitochondrial components, such as cytochromes b_1 and $a + a_2$, it has been suggested that these mutants might be controlled by the mitochondrial genome, but this has not been proved.

The mycelial variant of *Aspergillus nidulans* This example, which was studied before modern concepts of organelle genetics had developed, is one involving a cytoplasmic factor whose expression could be suppressed by certain nuclear genes (Roper, 1958). These interactions between the nuclear and cytoplasmic genomes have their parallels in other examples described elsewhere in this book (pp. 51, 109, 110).

The variants, denoted M_1, M_2 and M_3, appeared in cultures which had been treated with acriflavine, and differed from the normal strain in failure to produce condia or perithecia (except after prolonged growth). The genetic analysis was in three parts: (a) heterokaryons between mycelial and normal strains were prepared, and the homokaryons subsequently obtained showed recombination between the mycelial character and the nuclear genes, thus indicating cytoplasmic determination of mycelial; (b) crosses were made between mycelial and normal strains, and gave results indicating that either of two recessive genes m_1 and m_2 could produce the mycelial phenotype, and (c) heterokaryons between a non-mycelial strain of genotype m_1^+ m_2^+ derived from the cross in (b), and another non-mycelial strain containing nuclear marker genes (such as *paba-1*) were prepared, and amongst the homokaryons subsequently obtained, some were mycelial, thus showing that the cytoplasmic mycelial factor had been present in the apparently normal $m_1^+m_2^+$ strain. These genes, when present together, act to suppress the action of the cytoplasmic factor.

Some other variants of *A. nidulans*, denoted 'fluffy', showing some similarities to mycelial in their inheritance, have been described though in these cases the cytoplasmic factors differed from those in mycelial in that the 'fluffy' factors showed a capacity for invading normal strains.

A cytoplasmic suppressor (*psi-*) of supersuppressor genes in yeast In yeast the supersuppressor denoted *SUQ5* is a dominant nuclear gene which in combination with an adenine-requiring allele ($ad_{2,1}$) produces a pseudo-wild phenotype. The gene *SUQ5* also suppresses the mutant effect of some other genes ($bi_{5,2}$ and $ly_{1,1}$). The cytoplasmic factor *psi-* annuls the action of the supersuppressor *SUQ5* (Cox, 1965; Young & Cox, 1972). Since the colour of the wild-type yeast colonies on an appropriate medium is white and of the adenine-requiring mutant is red, presence of *SUQ5* and *psi-* can be noted from the colour of the colonies, as shown in Table 6.1.

The mutant *psi-* arose spontaneously several times and remained stable following intercrossing between different *psi-* strains. Genetic analysis by crossing the *psi-* type to the unsuppressed adenine-requiring strain shows that the supersuppressor gene *SUQ5*, though itself suppressed, is still present. The factor *psi-* shows non-Mendelian inheritance. Following crosses with normal strains, *psi-* usually 'disappears', though occasionally it is carried on. In that case the ascospore ratios are 4:0 *psi-*:normal, i.e. there is no genic segregation. By various crosses involving other genes which can be suppressed by *SUQ5*, it was shown that *psi-* controls the expression of *SUQ5*, rather than $ad_{2,1}$ or the other suppressible genes directly.

Studies on the inheritance of *psi-* in combination with the mitochondrial factors $\rho -$ (petite) and *ery*[r] (erythromycin-resistance) show that there is no effect of segregation of these mitochondrial factors on the behavior of *psi-*. Hence *psi-* is probably not on the mitochondrial DNA; if it is, it must be at a site remote from the regions concerned with respiratory activities or erythromycin resistance. The cellular location of *psi-*, its nature and mode of action are thus unknown.

Table 6.1 Colours of yeast colonies of various constitutions (after Cox, 1965)

Strain	Symbol	Colour
Wild type	+	White
Adenine requiring	$ad_{2,1}$	Red
Adenine requiring with supersuppressor	$ad_{2,1} + SUQ5$	White
Adenine requiring + supersuppressor +*psi*-	$ad_{2,1} + SUQ5 + psi$-	Red

An episome-like factor in yeast Some yeast mutants conferring resistance to the drugs venturicidin (VEN) and triethyltin (TET) have been found to show non-Mendelian inheritance, but are unlinked to any factors known to be located on mitochondrial DNA (Griffiths *et al.*, 1975). The VEN-resistant strains can be derived from ρ^0 petites, i.e. those lacking mitochondrial DNA altogether. The precise genetic basis of these mutants, which are apparently quite distinct from the mitochondrially inherited VEN and TET mutants described earlier (pp. 22–3), is not clear.

Another group of yeast workers (Guerineau *et al.*, 1974) have also described evidence for some non-mitochondrial (and non-nuclear) factors controlling resistance to various drugs, including VEN. Some of these effects seem to be correlated with the presence of a small, circular DNA molecule, distinct from mitochondrial DNA. It is possible that these extramitochondrial factors are plasmids or episomes, but there is no clear evidence for this at the present time.

Conclusion Many other examples of cytoplasmic heredity in fungi have been described, with varying amounts of detail. Even as early as 1927, Harder reported some experiments with the basidiomycetes *Pholiota* and *Schizophyllum*, which produce dikaryotic mycelia after fusion of the gametes (Harder, 1927a & b). By micromanipulation (cutting the clamp connections) Harder produced monokaryons which were presumed to contain mixed cytoplasm from the two different strains used in a cross, and certain growth habits were thought to be controlled by the cytoplasm. However, detailed genetic analysis with nuclear marker genes was not possible at that time, and Harder's hypothesis has been criticized by later workers (Jinks, 1966; Esser & Kuenen, 1967).

The examples described in this chapter illustrate the diversity of extranuclear genetic phenomena in fungi. As for the cellular determinants of the characters described, one can only speculate. Where there is easy infectibility, as in the case of the barrage phenomenon in *Podospora*, it is possible that viruses are involved. It has been suggested that many cytoplasmically inherited characters in fungi may have a viral basis, since a number of fungi contain viruses, though most of those known have pathological effects (Hollings & Stone, 1969). No doubt some of the examples will turn out to have a mitochondrial basis, and this seems likely for the mycelial variant of *Aspergillus*. The different types of senescence phenomena are the most difficult to relate to any particular cytoplasmic particle and it may be that these are of a non-genetic developmental nature.

A wide variety of mechanisms is possible, and it would be very interesting to devise experiments specifically to identify the factors concerned.

Cytoplasmic male sterility in plants

Introduction More than eighty species of angiosperms have been recorded as producing strains with cytoplasmically inherited pollen sterility (Edwardson, 1970). This is a matter of considerable economic importance since in maize and other agriculturally important crops such plants can be used for the production of hybrids, without the labour of emasculating the flowers of the plants to be used as female parents. Male sterility was first analysed genetically in flax (*Linum usitatissimum*) by Bateson and Gairdner (1921) and Chittenden and Pellew (1927). Two strains, procumbent and tall, were crossed. When the procumbent strain was used as the female parent, male-sterile plants appeared in one quarter of the plants of the F_2 generation, and it was concluded that male sterility was due to an interaction between a recessive gene from the tall parent and a cytoplasmic factor from the procumbent parent. The analysis was continued by von Wettstein (1946), who made a series of eight backcrosses to the tall strain and found that the cytoplasm from the procumbent strain remained unaltered in regard to its effect on pollen development.

Male sterility also occurs in *Oenothera* and *Epilobium*, two plants with which much study of cytoplasmic heredity has been made. This work led to long controversies in the German literature on the relative importance of the 'plastidom' or 'plastom' (genetic constituents of plastids) and the 'plasmon' (genetic factors not associated with any known particulate structure). The methods used, basically those of Mendel, were not sufficiently subtle to resolve these controversies, and as will be seen, even today the material basis of cytoplasmic male sterility has not been conclusively identified.

Genetics of cytoplasmic male sterility in maize Male sterility in maize was shown by Rhoades (1933) to be inherited through the egg cytoplasm. Using plants grown from seed of a male-sterile plant found in Peru, he made a series of crosses, using the male-sterile plants as female parents, with strains containing genes marking each of the ten chromosomes of maize. After all the chromosomes of the original male-sterile line had been replaced by chromosomes from the fertile line, many plants were still male-sterile. It was therefore concluded that male sterility was controlled by a cytoplasmic factor (*cms*). Duvick (1965) showed that such cytoplasmic transmission could be continued for more than twenty generations.

However, alongside the male-sterile plants in the progeny of the above crosses, some partially or completely fertile plants were obtained, and these were shown to contain certain 'restorer' genes, derived from the fertile lines. These are dominant nuclear genes, which when brought into the cytoplasm derived from a male sterile line, have the effect of restoring fertility. Some restorer genes (Rf_1, Rf_2, Rf_3) have been assigned to particular chromosomal loci. They do not permanently modify or eliminate the *cms* factors, but merely suppress their expression, for if crosses are made whereby the restorers are removed, i.e. replaced by their non-restoring alleles, the action of the male-sterile cytoplasm is re-established.

In different strains of maize showing cytoplasmic male sterility, different types of cytoplasmic factor may be present, as shown by their interactions with various restorer genes. In Table 6.2 the interactions of two cytoplasmic types called 'Texas' (or *cms-T*) and 'USDA' (or *cms-S*) with restorer genes in various inbred lines are shown. Line K55 contains a gene which restores pollen fertility to the Texas type; line CE1 restores the USDA type; line Ky 21 contains both types of restorers but line A158 neither. Further cytoplasmic types and further restorer genes are also known (Beckett, 1971). An interesting distinction between the time of action of the restorers Rf_1 and Rf_3 is that the former interacts with the *cms-S* cytoplasm in the diploid (sporophyte) generation, while the latter interacts with *cms-T* only in the haploid (gametophyte) cells. This conclusion was drawn from observations showing that plants having the heterozygous genotype Rf_2 rf_1 and cytoplasm *cms-S* yield nearly 100% fertile pollen, whereas the combination of Rf^3 rf_3 with *cms-T* gives only about 50%.

Table 6.2 Effect of various combinations of cytoplasmic types and restorer genes on pollen sterility in maize (Duvick, 1965).

| Inbred lines | A158 | Ky 21 | K55 | CE1 |
Restorer genes	—	$Rf_1 + Rf_3$	Rf_1	Rf_3
Cytoplasmic Type				
USDA (*cms-S*)	Sterile	Fertile	Sterile	Fertile
Texas (*cms-T*)	Sterile	Fertile	Fertile	Sterile

Some fertility-restoring genes do not produce complete fertility, and environmental factors also have a modifying influence. Hot dry weather at about the time of flowering causes plants with *cms-T* and partial restorer genes to show a greater degree of sterility, while cool humid weather produces a higher percentage of fertile pollen (Duvick, 1965). Using partial restorers it is possible to show that there is no transmission of *cms* factors through the pollen tubes. After a series of backcrosses of partial restorers—used as male parents—to normal plants have been made, the fertility of the progeny is unaffected (Rhoades, 1933; Duvick, 1965). Moreover, as stated above, cytoplasm from normal plants used as male parents in crosses with cytoplasmic male sterile females, does not cause any modification of the *cms* factors transmitted through the egg, at least in maize.

Cytoplasmic male sterility in other plants As already stated, this phenomenon occurs in many genera and species of plant. The mechanism seems rather similar in many of them: there is inheritance through the egg cytoplasm and often an interaction with nuclear restorer genes. It seems possible that almost any plant could be bred to produce a cytoplasmic male sterile type, if sufficient work were done.

An apparently unusual type of male sterility has been described in the bean *Vicia faba*, in that here fertility restoration, once it has occurred, is irreversible. Moreover, permanent *spontaneous* reversion of certain male-sterile plants to fertility has also been found in this species (Bond *et al.*, 1966). In another plant also, pearl millet (*Pennisetum typhoides*), 'mutation' of a *cms* factor to fertility has been found (Clement, 1975).

Male sterility often arises in the progeny of crosses between different species or genera. For example, in wheat, the combination of cytoplasm of *Triticum timopheevi* with chromosomes of common wheat (*T. vulgaris*), obtained by repeated backcrossing, yields male-sterile plants. Moreover, *T. timopheevi*, which is normally male-fertile, contains three dominant restorer genes which act cumulatively when in the cytoplasm of common wheat (Miller *et al.*, 1974; Sage, 1972). Similarly the cytoplasms of *Aegilops candata* or *Ae. ovata*, combined with the chromosomes of common wheat, also give rise to male-sterile plants. Again in tobacco, cytoplasm of *Nicotiana langsdorffi* combined with chromosomal genes of *N. sanderae* produces male sterility, and restorer genes have been found here too (Smith, 1968). Similar results have been reported from interspecific crosses of *Oenothera* and *Epilobium* species (Michaelis, 1933; Stubbe, 1959, 1964). Such nucleo-cytoplasmic interactions may also affect the development of chloroplasts, as discussed elsewhere (p. 51). The occurrence of cytoplasmic male sterility in the progeny of interspecific hybrids may be interpreted to indicate that the harmonious interactions between cytoplasmic and nuclear factors within a given species may be disrupted when cytoplasm from one species is combined with nuclei of another.

Transmission of cytoplasmic male sterility by infection Many attempts have been made to transmit *cms* factors to plants by some means other than passage through the egg cytoplasm, e.g. by infection. Most of these experiments have given negative results, but some appear to have been successful. The interpretation is somewhat controversial.

Rhoades (1933) inoculated the juice expressed from tassels and upper internodes of cytoplasmic male-sterile maize plants into young seedlings of normal lines, but no transmission of male sterility was found, either to the inoculated plants themselves or to their progeny. Moreover, as pointed out by Duvick (1965), infection is unlikely to have occurred under natural conditions, through chance contacts or insects, since vast numbers of fertile and male-sterile maize plants are grown together and no sterile plants have been observed among the normals.

In wheat, Lacadena (1968) dissected embryos and endosperms from normal and cytoplasmic male sterile plants (derived from *T. timopheevi*) and combined normal embryos with endosperms from cytoplasmic male sterile plants (and vice-versa), but no influence of one or the other was demonstrated in regard to cytoplasmic male sterility. Negative results have also been reported from grafting experiments with various plants, such as *Nicotiana, Triticum, Beta, Capsicum, Crotularia* (van Marrewijk, 1970), and *Vicia faba* (Bond *et al.*, 1966). In *Petunia*, however, success in obtaining the transmission of *cms* factors across the boundary between a male-sterile stock and a male-fertile scion has been reported (Frankel, 1956, 1962; Edwardson & Corbett, 1961). In these experiments the grafts themselves showed no change from fertility to sterility, but seed from flowers on the grafts yielded some male-sterile plants. However, according to van Marrewijk (1970), male-sterile flowers occasionally appear on grafts not involving cytoplasmic male sterile stocks, and are possibly due to irregular manifestation of restorer genes under changing environments. Some success in graft transmission of *cms* factors has also been reported from experiments with sugar beet (Curtis, 1967) and tobacco (Corbett & Edwardson, 1964).

In view of the conflicting interpretations of this matter, which is crucial for our understanding of the *cms* factors, further work is required to establish conclusively whether infection occurs, using homozygous plants which have been thoroughly studied in regard to restorer genes and environmental effects, as well as the inheritance of the *cms* cytoplasmic factors.

Other effects of cytoplasmic male sterility factors, especially on mitochondria Characters other than pollen fertility are sometimes slightly abnormal in cytoplasmic male sterile plants. For example in maize there may be a slight reduction in height and leaf number. Fertility on the female side is usually unaffected in maize, though in *Epilobium* and the *Aegilops/Triticum* hybrids there may be reduced female fertility (Duvick, 1965).

In maize, susceptibility to the fungus *Helminthosporium maydis* (race T), causing Southern Leaf Blight disease, is greater in *cms-T* than in normal plants, while other *cms* types seem to have unaltered susceptibility to this fungus (Smith *et al.*, 1971). This increased susceptibility of *cms-T* plants is correlated with some mitochondrial differences. Mitochondria isolated from *cms-T* plants, when treated with the fungus pathotoxin, show a decrease in respiration, oxidative phosphorylation and electron transfer, and may swell up; while mitochondria from normal plants do not show these effects when treated in the same way. It has been suggested that the toxin has a binding-site on the inner mitochondrial membrane, which is altered in *cms-T* plants (Miller & Koepple, 1971; Flavell, 1975).

These results have been followed up by studies with callus-cultures of *cms-T* and normal cells (Gengenbach & Green, 1975). Growth of *cms-T* cells, but not of normal cells, was found to be inhibited by a given concentration of toxin, and following treatment of *cms-T* cells with a mild dose of toxin for several rounds of growth, stable resistant cells appeared, having mitochondria similar to those in normal cells. It is possible that this change from sensitivity to resistance to the toxin is due to a mitochondrial mutation, though other possibilities exist, such as a gene mutation producing a restorer allele.

Discussion Although cytoplasmic male sterility in plants is clearly controlled by cytoplasmic factors acting in conjunction with certain nuclear genes, these cytoplasmic factors have not yet been conclusively identified. They may be viruses or virus-like particles, or alternatively mitochondrial or plastid genetic elements, or finally it may be that no particulate structure is required at all and some system of alternative steady states would explain the phenomena.

The virus type of hypothesis, favoured by Atanasoff (1964), is not supported by any direct evidences of viruses, apart from some small inclusions (50–60 nm in diameter) which have been observed in areas of dense cytoplasm in root-tip cells of *cms-T* maize and which are absent from normal cells. In *Vicia faba* spherical bodies (c. 70 nm in diameter) have also been seen in cells in cytoplasmic male sterile plants but not in normal cells (Edwardson, 1962; Edwardson *et al.*, 1976). No proof of transmission of these particles by infection has been obtained. Moreover, attempts to 'cure' cytoplasmic male sterility by such treatments as heat, chemicals or ionizing radiation, which are effective in eliminating sigma virus from *Drosophila* or kappa particles from *Paramecium*, have not been successful.

Mitochondria are certainly candidates for the position of bearers of *cms* factors. As described above, mitochondria in certain cytoplasmic male sterile types differ from those in fertile plants. Recent studies with restriction endonucleases indicate that the mitochondrial DNA of *cms-T* maize differs from the normal (Levings *et al.*, 1976); while *cms-S* mitochondria contain two additional DNA fragments, possibly episomes (Pring *et al.*, 1977). Plastids have also been considered to be concerned with cytoplasmic male sterility. In *Epilobium hirsutum* (Essen strain) a plastid mutation affects pollen fertility, and in *Oenothera* various genome complexes when combined with plastids from *Oe. parviflora* produce non-germinating pollen (Edwardson, 1970).

According to one hypothesis, cytoplasmic male sterility is the consequence of an interaction between substances in the anthers and some components of mitochrondria or chloroplasts. This would account for the fact that the only cells seriously defective in cytoplasmic male sterile plants are in the anthers, while presumably the extranuclear genetic factors concerned are present in all cells (Flavell, 1974). A totally different hypothesis has been proposed (Heslop-Harrison, 1964), according to which cytoplasmic male sterility is not caused by any cytoplasmic genetic particles at all, but by the alternative steady states of a regulator–operon system. Such a scheme would hardly account for the apparently permanent stability of the cytoplasm in cytoplasmic male sterile plants, and additional assumptions would be needed to explain the action of the restorer genes.

The most plausible mechanism would seem to be one based on a cytoplasmic organelle found in all higher plants, such as a mitochondrion—a view expressed as early as 1950 by Rhoades; but in the absence of conclusive evidence, such a hypothesis remains speculative at present. Possibly different mechanisms operate in different plants.

Nucleo-cytoplasmic interactions in protozoa

Amoeba Since no sexual process is known to occur in amoebae, genetic work by the usual methods is not possible with these organisms. They are nevertheless very favourable materials for experiments involving the transfer of cellular components by micromanipulation. Nuclei from one strain (or species) can be injected into the enucleated cytoplasm of another strain and new nuclear–cytoplasmic hybrids constructed; and cytoplasmic components from one strain can be injected into another. Finally, nuclei from one strain can be introduced into nucleated cells of another, producing heterokaryons, and from these, homokaryons can be produced, either by cutting the dikaryotic cell into two pieces, or by allowing cell division to take place, producing two daughter cells each with one nucleus (Jeon, 1973).

The interpretation of the results of such experiments is, from a genetic point of view, not as clear as might appear, however. Nuclear marker genes are of course unavailable, and the absence of a technique for genetic analysis makes it impossible to determine whether any changes in cell properties produced by manipulation are of a genetic nature. Moreover, the characters available for study, such as minor quantitative variations in resistance to drugs and other external agents, and shape and size of cell components, are rather imprecise. Transplantation of nuclei between different 'species' of *Amoeba* usually succeeds in only a small percentage of cells, and when cytoplasm is transferred the effects produced might be due to the endosymbionts which are sometimes present (see p. 84).

Notwithstanding these and other possible sources of confusion, evidence has been obtained that some characters are controlled by nuclear and others by cytoplasmic factors. For example, it was found by studies of nuclear–cytoplasmic hybrids between *A. discoides* and *A. proteus* that resistance to streptomycin and some other drugs was controlled by cytoplasmic components. By fractionating the cytoplasmic material before injection, evidence was obtained indicating that the fraction responsible was that containing ribosomes and smaller particles (Hawkins, 1973). Another group of workers, however, found that resistance to streptomycin in various *A. proteus* strains was under exclusively nuclear control (Kalinina, 1969).

Some results indicate that differences between strains of *Amoeba*, even though associated with nuclei, are still not genetic. Heterokaryons were made by combining nuclei derived from strains differing in sensitivity to methionine and high temperature and subsequently re-establishing uninucleate cells. These were shown to differ from both original strains, and in some cases to show marked instability. Yudin (1973) considered that there was a mutual interaction between the two nuclei in the heterokaryotic cell, producing changes in behaviour which could continue for a period after the nuclei had been separated. It was also suggested that some apparent effects of the cytoplasm could be of the same nature, producing a temporary non-genetic change in nuclear behaviour.

Paramecium   In another protozoan, *Paramecium aurelia*, there is also evidence that changes in cell properties, some persisting for long periods, may occur in the absence of any change in the basic genetic material. As an example, the immobilization antigens of *P. aurelia* may be briefly considered. These antigens are proteins covering the entire surface of the paramecia, and vary in a complex way. Genetic analysis shows that they are controlled by a series of genes at different chromosomal loci, each coding for a particular antigen. Under different environmental conditions, and sometimes even without any environmental change, genes at different loci are called into expression. Usually only genes at one antigen-determining locus are expressed at a given time in a given cell, the remaining genes being 'unexpressed', i.e. they have no effect on the phenotype. The control over which genes are expressed and which are prevented from being expressed is still not understood. Some experiments showed that a cytoplasmic factor is involved, though the cytoplasm is itself also influenced by both genes and environment (Beale, 1954).

The point to be stressed in the present context is that these cell changes, which involve a switch from production of one protein to another and which may appear to be of a semi-permanent nature, do not involve any irreversible mutation or loss of a genetic element; all the cell types produced by a given genome are ultimately capable of giving rise to all the others.

Paramecium and other ciliates show a number of other quasi-hereditary cell variations. As examples, the system of alternative mating types, controlled in some cases by variations in the behaviour of the macronucleus, and the patterns of surface structures, which are to some degree controlled by the pattern in pre-existing cells (Sonneborn, 1974), may be mentioned These phenomena, though of great interest in regard to problems of cell differentiation, are not directly related to the main theme of this book and hence will not be discussed further here. It should however be pointed out that in unicellular organisms, especially those lacking a sexual stage, it is sometimes difficult to distinguish these cellular modifications from truly genetic changes, whether based on nuclear or extranuclear genes. This can only be done where a precise genetic analysis is possible.

7 Conclusion: Extranuclear Genetics and the Evolution of Cell Organelles

In the previous chapters we have surveyed various types of extranuclear genetic system. It is obvious that, unlike the nuclear system which is remarkably uniform throughout the whole range of eukaryotic organisms, extranuclear genetic systems are very diverse. Many of our examples are based on small, usually circular, strands of DNA, lacking histones and chromatin structure; but the size of these DNA circles, and the amount of genetic information they carry, vary a great deal. Transmission of extranuclear genetic factors from generation to generation is also very variable: in many cases they are transmitted exclusively via the cytoplasm of the female reproductive cells, but in some cases also through the male cells. Rarely, transmission may be brought about by infection, either artificially through the external medium, or by rapid spread between cells and tissues. Since many extranuclear genes show only maternal inheritance, cells containing mixtures of such genes, or extranuclear 'heterozygotes', are infrequent, though they certainly occur. Even where they do, extranuclear segregation and recombination are much less regular than the Mendelian processes which occur during the meiotic divisions and ensure the regular behaviour of nuclear genes. In some cases, however, as for example those of mitochondrial genes in yeast or chloroplast genes in *Chlamydomonas*, segregation and recombination, though not completely understood, are remarkably frequent.

As stated at the beginning of this book, the fraction of heredity controlled by extranuclear genes is very small by comparison with that controlled by the nuclear genes. However, the extranuclear fraction, or at least that part of it which is in mitochondria and plastids, is quite essential for maintaining life in its present form. In view of the universality of occurrence of these separate genetic systems of mitochondria and plastids and of the dual genetic control of these organelles by nuclear and extranuclear genomes, it is natural to ask whether there is any advantage in these systems, which as mentioned earlier (p. 41) seem, at least to a human observer, unnecessarily complicated. It must be admitted that a satisfactory answer to this question has not been given. If the genetic control of mitochondria and chloroplasts were exclusively nuclear, these organelles would be expected to have opportunities of flexible evolution due to mutation, recombination and selection; but since part of the genetic control is extranuclear and the extranuclear systems are probably less regularly variable than the nuclear ones, the extranuclear systems do not seem to confer an advantage in an evolutionary sense. One way of answering the question posed by the presence of the organelle genetic systems is to deny that they have any particular advantage and assume that they are mere remnants of the evolutionary process, destined eventually to disappear. This seems, however, unlikely, because there has been more than ample time, since the pre-Cambrian period, for them to disappear. They have not disappeared and so we must assume that there is a positive advantage in their maintenance. This leads us to the question of the evolution of cell organelles.

Obviously such discussion is purely speculative. These evolutionary processes took place many millions of years ago, when eukaryotes are presumed to have been arising from prokaryotes, and it is very hazardous to reconstruct the detailed course of these events on the basis of principles derived from forms of life now existing under quite different conditions; and no experimental verification of any hypothesis is possible. As Stanier (1970) points out, we are here engaged in a kind of 'metascience', the scientific equivalent of the metaphysics of the philosophers. Such arguments cannot lead to any final conclusions.

Hypotheses on the evolution of organelles can be placed in two categories, which may be denoted (1) symbiotic and (2) non-symbiotic. The symbiotic type of hypothesis is the more complex, but since it is the favourite at the present time, we will consider it first.

Versions of the symbiotic hypothesis were put forward long ago by Altmann (1890) for mitochondria (which he called bioblasts) and by Mereschkowski (1905; 1910) for chloroplasts, but at those times of course nothing was known about the genetic systems of organelles. With the rise of genetics, these views fell into disrepute, since all but a few geneticists believed that genetic material lay exclusively in the nucleus, which was thought to govern the development of all cell characters. After the realization that cell organelles contain some hereditary systems of their own, and especially after the discovery of DNA in cell organelles, the symbiotic hypothesis gained renewed popularity. Such a hypothesis has been expounded in great detail by Margulis (1970; 1975), who has attempted to develop an all-embracing theory, not only of the evolution of mitochondria and chloroplasts, but of the eukaryotic cell *in toto*.

It is assumed that the earliest forms of life were anaerobic prokaryotes, since the earth's atmosphere originally consisted largely of nitrogen. After the development of photosynthesis, which is thought to have arisen first in organisms resembling blue-green algae and photosynthetic bacteria, increasing amounts of oxygen were liberated into the atmosphere, leading to a situation in which aerobic respiration would be an advantage to organisms possessing it. Margulis proposes that the aerobic mechanism developed in some bacterium-like organism, which was later ingested by another kind of organism—a 'protoeukaryote'—which had already developed to some extent in the direction of the eukaryotic cell, but was apparently unable to accomplish the evolution of a respiratory system of its own. The association of the aerobic prokaryotic symbiont with an anaerobic protoeukaryotic host is then thought to have become stabilized and the endosymbiosis became obligatory, due, it is suggested, to the long association of the symbiont with its host and consequent shedding of 'redundant' genetic information from the symbiont. Selection might be thought to result in a 'relegation' of dispensible metabolic functions to the host. If the process took place as proposed, the ingested symbiont would of course contain its own DNA and this would persist in the contemporary mitochondrion, subject to some loss due to the 'relegation' of redundant genes. Thus this type of hypothesis succeeds in explaining the presence of DNA in mitochondria.

At another stage a second act of ingestion took place, according to the symbiotic hypothesis, whereby some protoeukaryotic cells acquired photosynthetic endosymbionts derived from blue-green algae, which had evolved a photosynthetic system separate from the cytoplasmic membrane. Margulis assumes that the evolution of both respiration and photosynthesis occurred in prokaryotes, and preceded the evolution of the eukaryotic cell, but did not take place directly in the line of organisms destined to become eukaryotes. Stanier (1970) and others, however, consider it probable that the chloroplast was formed by symbiosis *before* the mitochondrion. Before considering the various arguments for and against the symbiotic hypothesis, we will briefly describe the completely different proposal of Raff and Mahler (1972; 1975), who dispense altogether with the necessity for symbiosis.

Raff and Mahler suggest that the primeval protoeukaryotic cells had already evolved a system of aerobic respiration, which like that of bacteria was located in the surface membranes, before mitochondria were developed. As the protoeukaryotic cells became larger, invaginations of the surface membranes were produced, increasing the surface area and eventually forming closed vesicles with the metabolic machinery for respiration on the inner surface. These vesicles thus became mitochondria, but the difficulty then is to explain the origin of mitochondrial DNA. To do this, Raff and Mahler propose that a plasmid which happened to be at large in the cytoplasm of the protoeukaryotic cell was 'captured' and installed inside the vesicles. It was further supposed that this plasmid acquired the genes for the various components of the mitochondrial protein-synthesizing system (r-RNAs, t-RNAS, etc), and certain hydrophobic components of the respiratory system, by multiple chromosome-plasmid gene exchanges.

In addition to these contrasting hypotheses, numerous other suggestions to explain the evolution of organelles have been made. Reijnders (1975) proposed that nuclear and mitochondrial DNAs developed from originally identical prokaryotic DNAs which became 'compartmentalized', i.e. separated from the rest of the cell and each other, by surrounding themselves with membranes, inside which differentiation then took place in different directions. Such a hypothesis does not require postulation of either a symbiont or a plasmid. Bogorad (1975) proposed a similar type of scheme (denoted 'cluster clone' hypothesis) for the evolution of chloroplasts, and a number of schemes of a somewhat similar nature have been put forward (Uzzell & Spolsky, 1974; Meyer, 1973).

In assessing the merits of these different ideas, one should remember that none can be substantiated by experiments, and even the observation of structures at an apparently intermediate stage of development between a prokaryotic organism and an organelle, like the bacterium-like endo-symbionts in the amoeba *Pelomyxa palustris* which lacks mitochondria (see p. 84), should be interpreted cautiously. Similarly the success of Jeon (1973) in establishing an obligatory symbiosis between an infective bacterium and *Amoeba discoides* does not imply that such a symbiont has evolved into a mitochondrion; and even if one were able to continue the experiment further and demonstrate that the newly established symbiont could acquire the functions of a mitochondrion, that would still not prove that existing mitochondria had in fact evolved in the same way in pre-Cambrian times. Success in laboratory manipulation is no proof that evolution has proceeded in the same way as that engineered by an experimental biologist.

Supporters of the symbiotic hypothesis have used as evidence the resemblances between cell organelles and known prokaryotic organisms. Mitochondria are similar in size to many bacteria and contain a circular strand of naked DNA and ribosomes which resemble those of bacteria in being susceptible to certain drugs, such as chloramphenicol, and resistant to others, such as cycloheximide. Similarly chloroplasts show a number of similarities to blue-green algae (Taylor, 1974). As for chloroplast ribosomal RNA, it has been shown by studies using an electrophoretic 'finger-printing' technique which reveal details of nucleotide sequences (Zablen *et al.*, 1975), that the 16S r-RNA of *Euglena* chloroplasts has certain similarities to prokaryotic r-RNAs. Mitochondria and chloroplasts show many prokaryotic features. There are, however, a number of differences between organelles and the prokaryotic organisms which are considered to have evolved into organelles. Mitochondrial DNA is much smaller than bacterial DNA and is more nearly equivalent in size to the DNA of a small plasmid or virus than of a bacterium. The mechanism of genetic recombination in mitochondria, so far as it is understood, seems more like that of viruses than of bacteria. The sensitivity of mitochondrial DNA to acridine dyes and ethidium bromide is also reminiscent more of plasmid than of bacterial DNA (Taylor, 1974). Hence, the genetic elements of mitochondria seem more nearly related to those of viruses than of bacteria.

On the other hand, in some respects mitochondria show some *eukaryotic* affinities, e.g. the messenger RNA of mitochondria has a high poly-A content, a characteristic feature not of bacterial but of eukaryotic m-RNA (Reijnders, 1975). Further, some mitochondrial components are distinct from those of both prokaryotes and of parts of the eukaryotic cell other than organelles. The sedimentation values of mitochondrial ribosomes, at one time thought to be like those of bacteria, are now known to show wide variation in different types of organism (see Table 2.1). Mitochondrial r-RNA apparently lacks a 5S component, which is found in both bacterial and eukaryotic r-RNA, and the RNA polymerases of mitochondria are also thought to show distinctive features. For example, mitochondrial RNA polymerase, while it is distinct from the nuclear RNA polymerase in sensitivity to inhibitors, template specificity and molecular weight, resembles that of bacteriophage T_7 more than the corresponding *E. coli* polymerase (Jacob, 1973).

One could list a great many characteristics of organelles and compare them with corresponding characteristics of prokaryotic and eukaryotic cells, but the conclusions drawn from these comparisons are of a somewhat dubious nature. This is because the great majority of organelle characters are formed not by the action of organelle genes, but of nuclear genes. Hence many of the similarities between prokaryotic organisms and organelles are 'phenotypic' and not due to similarities in the genetic systems of organelle and of the supposedly related prokaryotic organism. Of course, once having wholeheartedly accepted the symbiotic hypothesis, one could then assume that a massive transfer of genetic control from organelle to nuclear DNA took place while leaving the prokaryote-like phenotype relatively unchanged, but that seems a very large assumption. As Uzzell and Spolsky (1974) point out, the necessity to assume such a transfer weakens the argument for the symbiotic hypothesis. It is possible to imagine how the transfer might occur: there could be an insertion of a substantial portion of the organelle DNA in the nucleus, or, alternatively, already existing nuclear genes could mutate to take over functions of organelle genes, which would then be lost in the absence of any selective need for their maintenance. But these are no more than imaginable possibilities. On the whole the differences between the genetic material in organelles and in known bacteria are so great as to render the symbiotic hypothesis unlikely.

The non-symbiotic hypothesis of Raff and Mahler also suffers from some rather implausible assumptions, especially concerning a plasmid with genes for certain organelle functions. If these genes, for r-RNAs, t-RNAs and some respiratory proteins, were originally in the nucleus, they would most likely be scattered widely over the genome. To assemble them in one segment of DNA prior to excision and insertion in a plasmid would require a considerable amount of manipulation, difficult even by the standards of contemporary genetic engineering. Although plasmids can in principle be constructed to contain virtually any kind of DNA, and some are known to contain, for example, genes for t-RNAs (Meyer, 1973), it is improbable that a plasmid containing the genes for an alternative protein-synthesizing system would evolve just to make possible the *in situ* synthesis of certain hydrophobic proteins inside an organelle, as Raff and Mahler suppose. The plasmid element of this hypothesis is its least credible part.

Both types of hypothesis, the symbiotic and the non-symbiotic, are therefore subject to almost insuperable objections. We can only suggest that the evolution of organelles was a long and very complicated process of which many details are unknown to us. It is a situation so open to different interpretations that one can draw what conclusions one wishes. The raw material for making hypotheses of this kind is present in this book, and we will end it by recommending those temperamentally suited to this kind of mental activity to construct their own hypotheses.

References

ACHTMAN, M. (1973). Genetics of the F sex factor in Enterobacteriaceae. *Curr. Top. Microbiol. Immunol.*, **60**, 79–123.

ADAMS, G. M. W., VANWINKLE-SWIFT, K. P., GILLHAM, N. W. & BOYNTON, J. E. (1976). Plastid inheritance in *Chlamydomonas reinhardi*. In: *The Genetics of Algae*, ed. R. A. Lewin. Blackwell, Oxford, London, Edinburgh, Melbourne, 69–118.

ADOUTTE, A. (1977). La génétique des mitochondies chez la paramécie. Thèse d'état, Université de Paris sud.

ADOUTTE, A. & BEISSON, J. (1972). Evolution of mixed populations of genetically different mitochondria in *Paramecium aurelia*. *Nature*, **235**, 393–6.

AHMADJIAN, V. & HALE, M. E. (1973). *Lichens*. Academic Press, London.

ALEXANDER, N. J., GILLHAM, N. W. & BOYNTON, J. E. (1974). The mitochondrial genome of *Chlamydomonas*. *Molec. Gen. Genet.*, **130**, 275–90.

ALTMANN, R. (1890). *Die Elementorganismen und ihre Beziehungen zu den Zellen*. Publ. Veit, Leipsig, 1–145.

ANDERSON, E. S. & LEWIS, M. J. (1965). Characterization of a transfer factor associated with drug resistance in *Salmonella typhimurium*. *Nature*, **208**, 843–9.

AVANDHAMI, N. G. & BUETOW, D. E. (1972). Protein synthesis with isolated mitochondrial polysomes. *Biochem. Biophys. Res. Comm.*, **46**, 773–8.

ATANASOFF, D. (1964). Viruses and cytoplasmic heredity. *Z. Pfl. Zücht.*, **51**, 197–214.

ATTARDI, G., AMALRIC, F., CHING, E., COSTANTINO, P., GELFAND, R. & LYNCH, D. (1976). Informational content and gene mapping of mitochondrial DNA from HeLa cells. In: *The Genetic Function of Mitochondrial DNA*, eds. A. M. Kroon and C. Saccone. Elsevier North Holland, Amsterdam, 37–46.

BALL, G. B. (1968). Organisms living on and in protozoa. In: *Research in Protozoology*, *Vol. 3*, ed. T. T. Chen. Pergamon Press, Oxford and New York, 566–718.

BARATH, Z. & KÜNTZEL, H. (1972). Induction of mitochondrial RNA polymerase in *Neurospora crassa*. *Nature New Biol.*, **240**, 195–7.

BATESON, W. & GAIRDNER, A. E. (1921). Male-sterility in flax, subject to two types of segregation. *J. Genet.*, **11**, 269–75.

BEALE, G. H. (1954). *The Genetics of Paramecium aurelia*. Cambridge University Press, London, 1–178.

BEALE, G. H. (1969). A note on the inheritance of erythromycin resistance in *Paramecium aurelia*. *Genet. Res.*, **14**, 341–2.

BEALE, G. H. (1973). Genetic studies on mitochondrially inherited mikamycin-resistance in *Paramecium aurelia*. *Molec. Gen. Genet.*, **127**, 241–8.

BEALE, G. H., JURAND, A. & PREER, J. R. (1969). The classes of endosymbiont of *Paramecium aurelia*. *J. Cell Sci.*, **5**, 65–91.

BEALE, G. H. & KNOWLES, J. K. C. (1976). Interspecies transfer of mitochondria in *Paramecium aurelia*. *Molec. Gen. Genet.*, **143**, 197–201.

BEALE, G. H., KNOWLES, J. K. C. & TAIT, A. (1972). Mitochondrial genetics in *Paramecium*. *Nature*, **235**, 396–7.

BECKETT, J. B. (1971). Classification of male sterile cytoplasms in maize (*Zea mays*). *Crop Sci.*, **11**, 724–6.

BEISSON-SCHECROUN, J. (1962). Incompatibilité cellulaire et interactions nucléo-cytoplasmiques dans les phénomènes de 'barrage' chez le *Podospora anserina*. *Ann. Genet.*, **4**, 1–50.

BELCOUR, L. (1975). Cytoplasmic mutations isolated from protoplasts of *Podospora answerina*. *Genet. Res.*, **25**, 155–61.

BELCOUR, L. & BEGEL, O. (1977). Mitochondrial genes in *Podospora anserina*: recombination and linkage. *Molec. Gen. Genet.*, **153**, 11–21.

BELL, P. R. & MÜHLETHALER, K. (1964). The degeneration and reappearance of mitochondria in the egg cells of a plant. *J. Cell Biol.*, **20**, 235–48.

BERKALOFF, A., BREGLIANO, J. C. & OHANESSIAN, A. (1965). Mise en évidence de virions dans des Drosophiles infectées par le virus héréditaire *sigma*. *C.R. Acad. Sci.* (Paris), **260**, 5956–9.

BERNARDI, G., POUNELL, A., FONTY, G., KOPECKA, H. & STRAUSS, F. (1976). The mitochondrial genome of yeast: organization, evolution and the petite mutation. See Saccone & Kroon (1976), 185–98.

BERTRAND, H., MCDOUGALL, K. J. & PITTENGER, T. H. (1968). Somatic cell variation during uninterrupted growth of *Neurospora crassa* in continuous growth tubes. *J. Gen. Microbiol.*, **50**, 337–50.

BERTRAND, H. & PITTENGER, T. H. (1972a). Isolation and classification of extranuclear mutants of *Neurospora crassa*. *Genetics*, **71**, 521–33.

BERTRAND, H. & PITTENGER, T. H. (1972b). Complementation among cytoplasmic mutants of *Neurospora crassa*. *Molec. Gen. Genet.*, **117**, 82–90.

BERTRAND, H., SZAKACS, N. A., NARGANG, F. E., ZAGOZESKI, C. A., COLLINS, R. A. & HARRIGAN, J. C. (1976). The function of mitochondrial genes in *Neurospora crassa*. *Canadian J. Genet. Cytol.* **18**, 397–409.

BIGGER, A. H., MURRAY, K. & MURRAY, N. E. (1973). Recognition sequence of a restriction enzyme. *Nature New Biol.*, **244**, 7–10.

BIRNSTIEL, M. L., CHIPCHASE, M. & SPEIRS, J. (1971). The ribosomal RNA cistrons. In: *Progress in Nucleic Acid Research and Molecular Biology*, **11**, eds. J. N. Davidson and W. E. Cohn. Academic Press, New York and London, 351–89.

BLAIR, G. E. ELLIS, R. J. (1973). Protein synthesis in chloroplasts. I. Light-driven synthesis of the large sub-unit of fraction I protein by isolated pea chloroplasts. *Biochim. Biophys. Acta*, **319**, 223–34.

BLOSSEY, H.Ch. & KÜNTZEL, H. (1972). *In vitro* translation of mitochondrial DNA from *Neurospora crassa*. *FEBS Letters*, **24**, 335–8.

BOARDMAN, N. K. (1976). Chloroplast structure and development. In: *Harvesting the sun-photosynthesis in plant life*, eds. A. San Pietro, F. A. Greer and T J. Army. Academic Press, New York, 211–30.

BOARDMAN, N. K., LINNANE, A. W. & SMILLIE, R. M. (1971). *Autonomy and Biogenesis of Mitochondria and Plastids*. North Holland, Amsterdam and London.

BOGORAD, L. (1975). Evolution of organelles and eukaryotic genomes. *Science*, **188**, 891–8.

BOLOTIN, M., COEN, D., DEUTSCH, J., DUJON, B., NETTER, P., PETROCHILO, E. & SLONIMSKI, P. P. (1971). La recombinaison des mitochondries chez *Saccharomyces cerevisiae*. *Bull. Inst. Pasteur*, **69**, 215–39.

BOLOTIN-FUKUHARA, M., FAYE, G. & FUKUHARA, H. (1976). *Localization of some mitochondrial mutations in relation to transfer and ribosomal RNA genes in Saccharomyces cerevisiae*, eds. C. Saccone and A. M. Kroon. North Holland, Amsterdam, 243–50.

BOND, D. A., FYFE, J. L. & TOYNBEE-CLARKE, G. (1966). Male sterility in field beans (*Vicia faba*). III. Male sterility with a cytoplasmic type of inheritance. *J. Agric. Sci.*, **66**, 359–67.

BORST, P. (1972). Mitochondrial nucleic acids. *Ann. Rev. Biochem.*, **41**, 334–76.

BOYER, H. W. (1971). DNA restriction and modification mechanisms in bacteria. *Ann. Rev. Microbiol.*, **25**, 153–76.

BRUNS, G. A. P., EISENMAN, R. E. & GERALD, P. S. (1976). Human mitochondrial NADP-dependent isocitrate dehydrogenase in man-mouse somatic cell hybrids. *Cytogenetics & Cell Genetics*, **17**, 200–211.

BÜCHER, TH., NEUPERT, W., SEBALD, W. & WERNER, S. (Eds). (1976). *Genetics and*

124 References

biogenesis of chloroplasts and mitochondria. North Holland, Amsterdam.

BUNN, C. L., WALLACE, D. C. & EISENSTADT, J. M. (1974). Cytoplasmic inheritance of chloramphenicol resistance in mouse tissue culture cells. *Proc. Natl. Acad. Sci.*, **71**, 1681–5.

CABELLO, F., TIMMIS, K. & COHEN, S. N. (1976). Replication control in a composite plasmic constructed by *in vitro* linkage of two distinct replicons. *Nature*, 259, 285–90.

CAIRNS, J. (1963). The chromosome of *Escherichia coli*. *Cold Spring Harb. Symp.*, **27**, 43–6.

CAMPBELL, A. M. (1962). Episomes. *Adv. Genet.*, **11**, 101–45.

CHAN, P. H. & WILDMAN, S. G. (1972). Chloroplast DNA codes for the primary structure of the large subunit of Fraction I protein. *Biochim. Biophys. Acta*, **277**, 677–80.

CHARLES, H. P. & KNIGHT, B. C. J. G. (1970). Organization and control in prokaryotic and eukaryotic cells. *Symp. Soc. Gen. Microbiol.*, 20, Cambridge University Press, London.

CHATTON, É. (1925). *Pansporella perplexa. Ann. Sci. Nat. Zool.*, **8**, 5–84.

CHEVAUGEON, J. & DIGBEU, S. (1960). Un second facteur cytoplasmique infectant chez le *Pestalozzia annulata. C. R. Acad. Sci.* (Paris), **251**, 3043–5.

CHIANG, K.-S. (1968). Physical conservation of parental cytoplasmic DNA through meiosis in *Chlamydomonas reinhardi. Proc. Natl. Acad. Sci.*, **60**, 194–200.

CHITTENDEN, R. J. & PELLEW, C. (1927). A suggested interpretation of certain cases of anisogeny. *Nature*, **119**, 10–12.

CHIU, N., CHIU, A. & SUYAMA, Y. (1975). Native and imported transfer RNA in mitochondria. *J. Mol. Biol.*, **99**, 37–50.

CLARK-WALKER, G. D. & GABOR MIKLOS, G. L. (1974). Mitochondrial genetics, circular DNA and the mechanism of the *petite* mutation in yeast. *Genet. Res.*, **24**, 43–57.

CLARK-WALKER, G. D. & LINNANE, A. W. (1967). The biogenesis of mitochondria in *Saccharomyces cerevisiae*. A comparison between cytoplasmic respiratory-deficient mutant yeast and chloramphenicol-inhibited wild type cells. *J. Cell Biol.* **34**, 1–14.

CLEMENT, W. M. (1975). Plasmon mutations in cytoplasmic male-sterile pearl millet, *Pennisetum typhoides. Genetics, 79*, 583–8.

CLEWELL, D. B. & HELINSKI, D. R. (1969). Supercoiled circular DNA-protein complex in *Escherichia coli*: purification and induced conversion to an open circular DNA form. *Proc. Natl. Acad. Sci.*, **62**, 1159–66.

CLOWES, R. C. (1972). Molecular structure of bacterial plasmids. *Bact. Rev.*, **36**, 361–405.

COEN, D., DEUTSCH, J., NETTER, P., PETROCHILO, E. & SLONIMSKI, P. P. (1970). Mitochondrial genetics. I. Methodology and phenomenology. In: *Control of organelle development*, ed. P. L. Miller. *Soc. Exp. Biol. Symp.*, **24**, 449–96. Cambridge University Press, London.

COLLINS, J. & PRITCHARD, R. H. (1973). Relationship between chromosome replication and Flac episome replication in *E. coli. J. Mol. Biol.*, **78**, 143–56.

CORBETT, M. K. & EDWARDSON, J. R. (1964). Intergraft transmission of cytoplasmic male sterility. *Nature*, **201**, 847–8.

COX, B. S. (1965). Ψ, a cytoplasmic suppressor of super-suppressor in yeast. *Heredity*, **20**, 505–21.

CRAIG, I., TOLLEY, E. & BOBROW, M. (1976). Mitochondrial and cytoplasmic forms of fumarate hydratase assigned to chromosome 1. *Cytogenetics & Cell Genetics* **16**, 118–21.

CUMMINGS, D. J., GODDARD, J. M. & MAKI, R. A. (1976). *The Genetic Function of Mitochondrial DNA*, eds. G. Saccone and A. M. Kroon. North Holland, Amsterdam, 119–130.

CURTIS, G. J. (1967). Graft transmission of male sterility in sugar beet (*Beta vulgaris*). *Euphytica*, **16**, 419–24.

CURTISS, R. (1969). Bacterial conjugation. *Ann. Rev. Microbiol.*, **23**, 70–136.

DATTA, N. (1975). Epidemiology and Classification of Plasmids. In: *Microbiology 1974*, ed. D. Schessinger. American Society for Microbiology, Washington DC, 9–15.

DAVIDSON, R. G. & CORTNER, J. A. (1967). Mitochondrial malate dehydrogenase. A new genetic polymorphism in man. *Science*, **157**, 1569–71.

DAVIDSON, N., DEONIER, R. C., HU, S. & OHTSUBO, E. (1975). Electron microscope heteroduplex studies of sequence relations among plasmids of *E. coli*. X. DNA sequence organization of F and of F-primes, and the sequences involved in Hfr formation. In: *Microbiology 1974*, ed. D. Schlessinger. American Society for Microbiology, Washington DC, 56–65.

DAWID, I. B. (1972). Mitochondrial protein synthesis. In: *Mitochondria : Biogenesis and Bioenergetics*, ed. S. G. Vanden Bergh *et al*. Elsevier/North Holland, Amsterdam, 35–51.

DAWID, I. B. & BLACKLER, A. W. (1972). Maternal and cytoplasmic inheritance of mitochondrial DNA in *Xenopus*. *Devel. Biol.*, **29**, 152–61.

DAWID, I. B., HORAK, I. & COON, H. G. (1974). The use of hybrid somatic cells as an approach to mitochondrial genetics in animals. *Genetics*, **78**, 429–71.

DAWID, I. B., KLUKAS, C. K., OHI, S., RAMIREZ, J. L. & UPHOLT, W. B. (1976). Structure and evolution of animal mitochondrial DNA. See Saccone & Kroon (1976), 3–14.

DEMPSEY, W. B. & WILLETTS, N. S. (1976). Plasmid co-integrates of prophage lambda and R factor R100. *J. Bact.* **126**, 166–76.

DIACUMAKOS, E. G., GARNJOBST, L. & TATUM, E. L. (1965). A cytoplasmic character in *Neurospora crassa*. The role of nuclei and mitochondria. *J. Cell Biol.*, **26**, 427–43.

DILTS, J. (1976). Covalently closed, circular DNA in kappa endosymbionts of Paramecium. *Genet. Res.*, **27**, 161–70.

DOUGHERTY, E. C. (1957). Neologisms needed for structures of primitive organisms. *J. Protozool.*, **4**, (suppl.) 14.

DUJON, B., SLONIMSKI, P. P. & WEILL, L. (1974). Mitochondrial genetics. IX. A model for recombination and segregation of mitochondrial genomes in *Saccharomyces cerevisiae*. *Genetics*, **78**, 415–37.

DUVICK, D. N. (1965). Cytoplasmic pollen sterility in corn. *Adv. Genet.*, **13**, 2–56.

EDWARDSON, J. R. (1956). Cytoplasmic male-sterility. *Bot. Rev.*, **22**, 696–738.

EDWARDSON, J. R. (1962). Cytoplasmic differences in T-type cytoplasmic male sterile corn and its maintainer. *Amer. J. Bot.*, **49**, 184–7.

EDWARDSON, J. R. (1970). Cytoplasmic male-sterility. *Bot. Rev.*, **36**, 341–420.

EDWARDSON, J. R., BOND, D. A. & CHRISTIE, R. G. (1976). Cytoplasmic sterility factors in *Vicia faba*. *Genetics*, **82**, 443–9.

EDWARDSON, J. R. & CORBETT, M. K. (1961). Asexual transmission of cytoplasmic male sterility. *Proc. Natl. Acad. Sci.*, **47**, 390–6.

ELLIS, R. J. (1975). The synthesis of chloroplast membranes in *Pisum sativum*. In: *Membrane Biogenesis*, ed. A. Tzagoloff. Plenum, New York, 247–78.

ELLIS, R. J. (1976). The synthesis of chloroplast proteins. In: *Nucleic acids and protein synthesis in plants*, ed. R. Baker. Plenum, London.

ELLIS, R. J. & HARTLEY, M. R. (1974). Nucleic acids of chloroplasts. In: *Nucleic Acids*, ed. K. Burton, *Biochemistry*, 6, Medical & Technical Publ. Co., Lancaster and Butterworth, London, 323–49.

EPHRUSSI, B. (1953). *Nucleo-cytoplasmic relations in micro-organisms*. Clarendon Press, Oxford.

ESSER, K. & KUENEN, R. (1967). *Genetics of Fungi*, Springer, Berlin, Heidelberg and New York.

FALKOW, S. (1975). *Infectious multiple drug resistance*, Pion, London.

FALKOW, S., WOHLHIETER, J. A., CITARELLA, R. V. & BARON, L. S. (1964). Transfer of episomic elements to *Proteus*. *J. Bact.*, **87**, 209–19.

FAURÉ-FREMIET, E. (1952). Symbiontes bactériens des ciliés du genre *Euplotes*. *C.R.*

Acad. Sci. (Paris), **235**, 402–3.

FAYE, G. *et al.* (1973). Mitochondrial nucleic acids in the petite colonie mutants: deletions and repetitions of genes. *Biochimie*, **55**, 779–92.

FAYE, G., KUJAWA, C. & FUKUHARA, H. (1974). Physical and genetic organization of petite and grande yeast mitochondrial DNA. *J. Mol. Biol.*, **88**, 185–203.

FAYE, G., KUJAWA, C., DUJON, B., BOLOTIN-FUKUHARA, M., WOLF, K., FUKUHARA, H. & SLONIMISKI, P. P. (1975). Localization of the gene coding for the mitochondrial 16S ribosomal RNA using rho-mutants of *Saccharomyces cerevisiae*. *J. Mol. Biol.*, **99**, 203–17.

FINCHAM, J. R. S. & DAY, P. R. (1971). *Fungal Genetics*. 3rd Ed., Blackwell, Oxford and Edinburgh.

FLAVELL, R. (1974). A model for the mechanism of cytoplasmic male sterility in plants, with special reference to maize. *Plant Sci. Letters*, **3**, 259–63.

FLAVELL, R. (1975). Inhibition of electron transport in maize mitochondria by *Helminthosporium maydis* race T pathotoxin. *Physiol. Plant Path.*, **6**, 107–16.

FOSTER, T. J. & HOWE, T. G. B. (1971). Recombination and complementation between R factors in *E. coli* K12. *Genet. Res.*, **18**, 287–97.

FOUTS, D. L., MANNING, J. E. & WOLSTENHOLME, D. R. (1975). Physico-chemical properties of kinetoplast DNA from *Crithidia acanthocephali*, *Crithidia lucilia*, and *Trypanosoma lewisi*. *J. Cell Biol.*, **67**, 378–99.

FRANKEL, R. (1956). Graft induced transmission of cytoplasmic male sterility in *Petunia*. *Science*, **124**, 684–5.

FRANKEL, R. (1962). Further evidence on graft induced transmission to progeny of cytoplasmic male sterility in *Petunia*. *Genetics*, **47**, 641–6.

FRÉDÉRICQ, P. (1957). Colicins. *Ann. Rev. Microbiol.*, **11**, 7–22.

FRÉDÉRICQ, P. (1963). On the nature of colicinogenic factors: a review. *J. Theor. Biol.* **4**, 159–65.

FUKUHARA, H., FAYE, G., MICHEL, F., LAZOWSKA, J., DEUTSCH, J., BOLOTIN-FUKUHARA, M. & SLONIMISKI, P. P. (1974). Physical and genetic organization of petite and grande yeast mitochondrial DNA. *Molec. Gen. Genet.*, **130**, 215–38.

FUKUHARA, H., BOLOTIN-FUKUHARA, M., HSU, H. J. & RABINOWITZ, M. (1976). Deletion mapping of mitochondrial transfer RNA genes in *Saccharomyces cerevisiae* by means of cytoplasmic petite mutants. *Molec. Gen. Genet.*, **145**, 7–17.

GELVIN, S., HEIZMANN, P. & HOWELL, S. H. (1977). Identification and cloning of the chloroplast gene coding for the large subunit of ribulose-1,5-bisphosphate carboxylase from *Chlamydomonas reinhardi*. *Proc. Natl. Acad. Sci.* **74**, 3193–7.

GENGENBACH, B. G. & GREEN, C. E. (1975). Selection of T-cytoplasm maize callus cultures resistant to *Helminthosporium maydis* race T pathotoxin. *Crop Sci.*, **15**, 645–9.

GIBSON, I. (1973). Transplantation of killer endosymbionts of *Paramecium*. *Nature*, **241**, 127–9.

GILBERT, W. & DRESSLER, D. (1968). DNA replication: the rolling circle model. *Cold Spring Harbor Symp.*, **33**, 473–84.

GILLHAM, N. W. (1974). Genetic analysis of the chloroplast and mitochondrial genomes. *Ann. Rev. Genet.*, **8**, 347–91.

GILLHAM, N. W., BOYNTON, J. E. & LEE, R. W. (1974). Segregation and recombination of non-Mendelian genes in *Chlamydomonas*. *Genetics*, **78**, 439–57.

GLOVER, D. M., WHITE, R. L., FINNEGAN, D. J. & HOGNESS, D. S. (1975). Characterization of six cloned DNAs from *Drosophila melanogaster*, including one that contains the genes for rRNA. *Cell*, **5**, 149–57.

GODDARD, J. M. & CUMMINGS, D. J. (1975). Structure and replication of mitochondrial DNA from *Paramecium aurelia*. *J. Mol. Biol.*, **97**, 593–609.

GOWDRIDGE, B. (1956). Heterokaryons between strains of *Neurospora crassa* with different cytoplasms. *Genetics*, **41**, 780–9.

GREEN, P. B. (1964). Cinematic observations on the growth and division of chloroplasts in *Nitella*. *Amer. J. Bot.*, **51**, 334–42.

GRIFFITHS, D. E. & HOUGHTON, R. L. (1974). Studies on energy-linked reactions: modified mitochondrial ATPase of oligomycin-resistant mutants of *Saccharomyces cerevisiae*. *Eur. J. Biochem.*, **46**, 157–67.

GRIFFITHS, D. E., LANCASHIRE, W. E. & ZANDERS, E. D. (1975). Evidence for an extrachromosomal element involved in mitochondrial function: a mitochondrial episome? *FEBS Letters*, **53**, 126–30.

GUERINEAU, M., SLONIMSKY, P. P. & AVNER, P. (1974). Yeast episome: oligomycin resistance associated with a small covalently closed non-mitochondrial circular DNA. *Biochem. Biophys. Res. Comm.*, **61**, 462–9.

HALL, R. M., NAGLEY, P. & LINNANE, A. W. (1976). Biogenesis of mitochondria, XLII. Genetic analysis of the control of cellular mitochondrial DNA levels in *Saccharomyces cerevisiae*. *Molec. Gen. Genet.* **145**, 169–175.

HANSON, M. R., DAVIDSON, J. N., METS, L. J. & BOGORAD, L. (1974). Characterization of chloroplast and cytoplasmic ribosomal proteins of *Chlamydomonas reinhardi* by two-dimensional gel electrophoresis. *Molec. Gen. Genet.*, **132**, 105–18.

HARDER, R. (1972a). Zur Frage nach der Rolle von Kern und Protoplasma im Zellgeschehen und bei der Übertragung von Eigenschaften. *Zeitsch. Botan.*, **19**, 337–407.

HARDER, R. (1927b). Über mikrochirurgischen Operationen an Hymenomyceten. *Z. Wiss. Mikr.*, **44**, 173–82.

HARDY. K. G. (1975). Colicinogeny and related phenomena. *Bact. Rev.*, **39**, 464–515.

HARTLEY, M. R., WHEELER, A. & ELLIS, R. J. (1975). Protein synthesis in chloroplasts. V. Translation of messenger RNA for the large subunit of Fraction I protein in a heterologous cell-free system. *J. Mol. Biol.*, **91**, 67–77.

HASHIMOTO, M. & HIROTA, Y. (1966). Gene recombination and segregation of resistance factor R in *Escherichia coli*. *J. Bact.*, **91**, 51–62.

HAWKINS, S. E. (1973). Genetic information in the cytoplasm of Amoebae. In: *The Biology of Amoeba*, ed. K. W. Jeon. Academic Press, New York and London, 525–47.

HAWKINS, S. E. & WOLSTENHOLME, D. R. (1967). Cytoplasmic DNA-containing bodies and the inheritance of certain cytoplasmically determined characters in *Amoeba*. *Nature*, **210**, 928–9.

HAYES, W. (1968). *The Genetics of Bacteria and their Viruses* (2nd ed.), Blackwell, Oxford and Edinburgh.

HECKMAN, K. (1975). *Omikron*, ein essentieller Endosymbiont von *Euplotes aediculatus*. *J. Protozool.*, **22**, 97–104.

HEDGEPETH, J., GOODMAN, H. M. & BOYER, H. (1972). DNA nucleotide sequence recognized by the R1 endonuclease. *Proc. Natl. Acad. Sci.*, **69**, 3448–52.

HELINSKI, D. R., LOVETT, M. A., WILLIAMS, P. H., KATZ, L., KUPERSZTOCH-PORTNOY, Y., GUINEY, D. G. & BLAIR, D. G. (1975). Plasmid DNA Replication. In: *Microbiology 1974*, ed. D. Schlessinger. American Society for Microbiology, Washington DC, 104–14.

HERRING, A. H. & BEVAN, E. A. (1974). Virus-like particles associated with the double-stranded RNA species found in killer and sensitive strains of the yeast *Saccharomyces cerevisiae*. *J. Gen. Virol.*, **22**, 387–94.

HERRMANN, R. G. (1973). Number and arrangement of genomes in chloroplasts. *Genetics* (Suppl.), **74**, S114.

HERSHFIELD, V., BOYER, H. W., YANOFSKY, C., LOVETT, M. & HELINSKI, D. R. (1974). Plasmid Col E1 as a molecular vehicle for cloning and amplification of DNA. *Proc. Natl. Acad. Sci.*, **71**, 3455–9.

HERSHFIELD, V., BOYER, H. W., CHOW, L. & HELINSKI, D. R. (1976). Characterization of a mini-Col E1 plasmid. *J. Bact.*, **126**, 447–53.

HERTIG, M. (1936). The rickettsia, *Wolbachia pipientis*, and associated inclusions of the mosquito, *Culex pipiens*. *Parasitology*, **28**, 453–86.

HESLOP-HARRISON, J. (1964). Sex expression in flowering plants. In: *Meristems and*

Differentiation, Brookhaven Symp. Biol., **16**. Brookhaven Laboratory, New York, 109–25.

HEYNINGEN, V. VAN, CRAIG, I. & BODMER, W. (1973). Genetic control of mitochondrial enzymes in human-mouse somatic cell hybrids. *Nature*, **232**, 509–12.

HEYNINGEN, V. VAN, BOBROW, M., BODMER, W. F., GARDINER, S. E., POVEY, S. & HOPKINSON, D. A. (1975). Chromosome assignment of some human enzyme loci. *Ann. Hum. Genet.*, **38**, 295–303.

HOFFMAN, H. P. & AVERS, C. J. (1973). Mitochondrion of yeast: ultrastructural evidence for one giant, branched organelle per cell. *Science*, **181**, 749–50.

HOLLIDAY, R. (1969). Errors in protein synthesis and clonal senescence in fungi. *Nature*, **221**, 1224–8.

HOLLINGS, M. & STONE, O. M. (1969). Viruses in fungi. *Sci. Prog. Oxf.*, **57**, 371–91.

HU, S., OHTSUBO, E. & DAVIDSON, N. (1975). Electron microscope heteroduplex studies of sequence relations among plasmids of *E. coli*: structure of F13 and related F-primes. *J. Bact.*, **122**, 749–63.

HUTCHISON, C. A., NEWBOLD, J. E., POTTER, S. S. & EDGELL, M. H. (1974). Maternal inheritance of mammalian mitochondrial DNA. *Nature*, **251**, 536–38.

IKEDA, H., INUZUKA, M. & TOMIZAWA, J. (1970). P1-like plasmid in *E. coli* 15. *J. Mol. Biol.*, **50**, 457–70.

JACOB, F., BRENNER, S. & CUZIN, F. (1963). On the regulation of DNA replication in bacteria. *Cold Spring Harbor Symp.*, **28**, 329–48.

JACOB, F. & WOLLMAN, E. L. (1958). Les épisomes, éléments génétiques ajoutés. *C.R. Acad. Sci. (Paris)*, **247**, 154–6.

JACOB, S. T. (1973). Mammalian RNA polymerases. In: *Progress in Nucleic Acid Research and Molecular Biology*, **13**, ed. J. N. Davidson and W. E. Cohn. Academic Press, New York and London, 93–126.

JACOB, S. T., SCHINDLER, P. G. & MORRIS, H. P. (1972). Mitochondrial polyriboadenylate polymerase: relative lack of activity in Hepatomas. *Science*, **178**, 639–40.

JASKUNAS, S. R., NOMURA, M. & DAVIES, J. (1974). Genetics of bacterial ribosomes. In: *Ribosomes*, eds. M. Nomura, A. Tissières and P. Lengyel. *Cold Spring Harbor Laboratory*, 333–68.

JEON, K. W. (ED.) (1973). *The Biology of Amoeba*, Academic Press, New York and London.

JINKS, J. L. (1959). Lethal, suppressive cytoplasms in aged clones of *Aspergillus glaucus*. *J. Gen. Microbiol.*, **21**, 397–409.

JINKS, J. L. (1964). *Extrachromosomal Inheritance*, Prentice-Hall, Englewood Cliffs, New Jersey.

JINKS, J. L. (1966). Extranuclear inheritance. In: *The Fungi, Vol. 2*, eds. G. C. Ainsworth and A. S. Sussman. Academic Press, New York and London, 619–60.

KAHN, P. L. (1968). Evolution of a site of specific homology on the chromosome of *E. coli*. *J. Bact.*, **100**, 269–75.

KALININA, L. V. (1969). Studies on resistance of *Amoeba* to the action of streptomycin (Russian). *Tsitologiya*, **11**, 760–7.

KARAKASHIAN, M. W. (1975). Symbiosis in *Paramecium bursaria*. In: *Symbiosis*, eds. D. H. Jennings and D. L. Lee. *Symp. Soc. Exp. Biol.*, **29**, Cambridge University Press, London, 145–74.

KASAMATSU, H. & VINOGRAD, J. (1974). Replication of circular DNA in eukaryotic cells. *Ann. Rev. Biochem.*, **43**, 695–719.

KAWASHIMA, N. & WILDMAN, S. G. (1972). Studies on Fraction I protein. IV Mode of inheritance of primary structure in relation to whether chloroplast or nuclear DNA contains the code for a chloroplast protein. *Biochim. Biophys. Acta*, **262**, 42–9.

KIRBY, H. (1941). Organisms living on and in protozoa. In: *Protozoa in Biological Research*, eds. G. N. Calkins and F. M. Summers. Columbia University Press, New York, 1009–113.

KIRK, J. T. O. (1971). Will the real chloroplast DNA please stand up. In: *Autonomy and Biogenesis of Mitochondria and Chloroplasts*, eds. N. K. Boardman, A. W. Linnane and R. M. Smillie. North Holland, Amsterdam, 267–76.

KIRK, J. T. O. & TILNEY-BASSETT, R. A. E. (1967). *The Plastids*, Freeman, London and San Francisco.

KLEISEN, C. M., WEISLOGEL, P. O., FONCK, K. & BORST, P. (1976). The structures of kinetoplast DNA. 2. Characterization of a novel component of high complexity present in the kinetoplast DNA network of *Crithidia luciliae*. *Eur. J. Biochem.*, **64**, 153–60.

KNOWLES, J. K. C. (1974). An improved microinjection technique in *Paramecium aurelia*. Transfer of mitochondria conferring erythromycin-resistance. *Exp. Cell Res.*, **88**, 79–87.

KNOWLES, J. K. C. & TAIT, A. (1972). A new method for studying the genetic control of specific mitochondrial proteins in *Paramecium aurelia*. *Molec. Gen. Genet.*, **117**, 53–9.

KOIZUMI, S. (1974). Microinjection and transfer of cytoplasm in *Paramecium*. *Exp. Cell Res.*, **88**, 74–8.

KOLODNER, R. D. & TEWARI, K. K. (1972). Circular mitochondrial DNA (70×10^6 daltons) from pea leaves. (Abstract). *Fed. Proc.*, **31**, 876.

KOLODNER, R. D. & TEWARI, K. K. (1975). Chloroplast DNA from higher plants replicates by both the Cairns and the rolling circle mechanism. *Nature*, **256**, 708–11.

KOLTIN, Y. & DAY, P. R. (1976). Inheritance of killer phenotypes and double-stranded RNA in *Ustilago maydis*. *Proc. Natl. Acad. Sci.*, **73**, 594–8.

KÜNTZEL, H., BARATH, Z., ALI, I., KIND, J. & ALTHAUS, H. H. (1973). Virus-like particles in an extranuclear mutant of *Neurospora crassa*. *Proc. Natl. Acad. Sci.*, **70**, 1574–8.

LACADENA, J. R. (1968). Hybrid wheat. VII. Tests on the transmission of cytoplasmic male sterility in wheat by embryo-endosperm grafting. *Euphytica*, **14**, 439–44.

LACEY, R. W. (1975). Transfer of chromosomal genes between staphylococci in mixed cultures. *J. Gen. Microbiol.*, **71**, 399–401.

LAMBOWITZ, A. M., CHUA, N.-H. & LUCK, D. J. L. (1976). Mitochondrial ribosome assembly in *Neurospora*. Preparation of mitochondrial ribosomal precurser particles, site of synthesis of mitochondrial ribosomal proteins and studies on the *poky* mutant. *J. Mol. Biol.*, **107**, 223–53.

LANCASHIRE, W. E. & GRIFFITHS, D. E. (1975*a*). Studies on energy-linked reactions: isolation, characterisation and genetic analysis of trialkyl-tin-resistant mutants of *Saccharomyces cerevisiae*. *Eur. J. Biochem.*, **51**, 377–92.

LANCASHIRE, W. E. & GRIFFITHS, D. E. (1975*b*). Studies on energy-linked reactions: Genetic analysis of venturicidin-resistant mutants. *Eur. J. Biochem.*, **51**, 403–13.

LAVEN, H. (1959). Speciation by cytoplasmic isolation in *Culex pipiens*. *Cold Spring Harbor Symp.*, **24**, 166–73.

LAVEN, H. (1967). Speciation and evolution in *Culex pipiens*. In: *Genetics of Insect Vectors of Diseases*, eds. J. W. Wright and R. Pal. Elsevier, Amsterdam, 251–75.

LEDERBERG, J. (1952). Cell genetics and hereditary symbiosis. *Physiol. Rev.*, **32**, 403–30.

LEE, R. W. & JONES, R. F. (1973). Induction of Mendelian and non-Mendelian streptomycin resistant mutants during the synchronous cell cycle of *Chlamydomonas reinhardtii*. *Molec. Gen. Genet.*, **121**, 99–108.

LEISTER, D. E. & DAWID, I. B. (1975). Mitochondrial ribosomal proteins in *Xenopus laevis x mulleri* interspecific hybrids. *J. Mol. Biol.*, **96**, 119–24.

LEMKE, P. A. (1976). Viruses of eukaryotic microorganisms. *Ann. Rev. Microbiol.*, 30, 105–46.

LEMKE, P. A. & NASH, C. H. (1975). Fungal viruses. *Bact. Rev.*, **38**, 29–56.

LEVENTHAL, E. (1968). The sex ratio condition in *Drosophila bifasciata*; its

LEVINGS, C. S., III & PRING, D. R. (1976). Restriction endonuclease analysis of mitochondrial DNA from normal and Texas cytoplasmic male-sterile maize. *Science*, **193**, 158–60.

experimental transmission to several species of *Drosophila*. *J. Invert. Path.*, **11**, 170–83.

L'HÉRITIER, PH. (1955). Les virus intégrés et l'unité cellulaire. *Ann. Biol.*, **31**, 483–96.

L'HÉRITIER, PH. (1958). The hereditary virus of *Drosophila*. *Adv. Virus Res.*, 5, 195–245.

L'HÉRITIER, PH. (1970). *Drosophila* viruses and their role as evolutionary factors. In: *Evolutionary Biology, Vol. 4*, eds. T. Dobzhansky, M. K. Hecht and W. C. Steeve. Appleton-Century-Crofts, New York, 185–209.

L'HÉRITIER, PH. & HUGON de SCOEUX, F. (1947). Transmission par greffe et injection de la sensibilité héréditaire au gaz carbonique chez la *Drosophila*. *Bull. Biol. France Belg.*, **81**, 70–91.

L'HÉRITIER, PH. & TEISSIER, G. (1937). Une anomalie physiologique héréditaire chez la *Drosophila*. *C.R. Acad. Sci. (Paris)*, **205**, 1099.

LINNANE, A. W., LUKINS, H. B., MOLLOY, P. L., NAGLEY, P., RYTKA, J. & SRIPRAKASH, K. S. (1976). Biogenesis of mitochondria: molecular mapping of the mitochondrial genome of yeast. *Proc. Natl. Acad. Sci.*, **73**, 2082–85.

LINNANE, A. W., SAUNDERS, G. W., GINGOLD, E. B. & LUKINS, H. B. (1968). The biogenesis of mitochondria. V. Cytoplasmic inheritance of erythromycin resistance in *Saccharomyces cerevisiae*. *Proc. Natl. Acad. Sci.*, **59**, 903–10.

LIZARDI, P. M. & LUCK, D. J. L. (1972). The intracellular site of synthesis of mitochondrial ribosomal proteins in *Neurospora crassa*. *J. Cell Biol.*, **54**, 56–74.

LOISEAUX, S., MACHE, R. & ROZIER, C. (1975). Rifampicin inhibition of the plastid rRNA synthesis of *Marchantia polymorpha*. *J. Cell Sci.*, **17**, 327–35.

LONGO, G. P. & SCANDALIOS, J. G. (1969). Nuclear gene control of mitochondrial malic dehydrogenase in maize. *Proc. Natl. Acad. Sci.*, **62**, 104–11.

LUKINS, H. B., TATE, W. R., SAUNDERS, G. W. & LINNANE, A. W. (1973). The biogenesis of mitochondria 26. *Molec. Gen. Genet.*, **120**, 17–25.

MCLENNAN, A. G. & KEIR, H. M. (1975). Subcellular location and growth stage dependence of the DNA polymerases of *Euglena gracilis*. *Biochim. Biophys. Acta.*, **407**, 253–62.

MANNING, J. E., WOLSTENHOLME, D. R. & RICHARDS, O. C. (1972). Circular DNA molecules associated with chloroplasts of spinach, *Spinacia oleracea*. *J. Cell Biol.*, **53**, 594–601.

MARCOU, D. (1961). Notion de longévité et nature cytoplasmique du déterminant de la sénescence chez quelques champignons. *Ann. Sci. Nat. Bot.*, *(12e série)*, **2**, 653–764.

MARGULIS, L. (1970). *Origin of eukaryotic cells*, Yale University Press, New Haven and London, 1–349.

MARGULIS, L. (1975). Symbiotic theory of the origin of eukaryotic organelles: criteria for proof. In: *Symbiosis*, eds. D. H. Jennings and D. L. Lee. *Symp. Soc. Exp. Biol.*, **29**, Cambridge University Press, Cambridge, 21–38.

MARREWIJK, G. A. M. VAN (1970). Cytoplasmic male sterility in *Petunia*. II. A discussion on male sterility transmission by means of grafting. *Euphytica*, **19** (1), 25–32.

MARVIN, D. A. & HOHN, B. (1969). Filamentous bacterial viruses. *Bact. Rev.*, **33**, 172–209.

MERESCHKOWSKI, C. (1905). Über Natur und Ursprung der Chromatophoren im Pflanzenreiche. *Biol. Zentr.*, **25**, 593–604.

MERESCHKOWSKI, C. (1910). Theorie der zwei Plasmaarten als Grundlage der Symbiogenesis, einer neuen Lehre der Entstehung der Organismen. *Biol. Zentr.*, **30**, 278–88.

MEYER, R. R. (1973). On the evolutionary origin of mitochondrial DNA. *J. Theor. Biol.*, **38**, 647–63.

MEYNELL, G. C. (1972). *Bacterial plasmids*, Macmillan, London.

MEYNELL, G. C. & LAWN, A. M. (1967). Sex pili and common pili in the conjugational transfer of colicin factor Ib by *Salmonella typhimurium*. *Genet. Res.*, **9**, 359–67.

MICHAELIS, G., PETROCHILO, E. & SLONIMSKI, P. P. (1973). Mitochondrial genetics. III. Recombined molecules of mitochondrial DNA obtained from crosses between cytoplasmic petite mutants of *Saccharomyces cerevisiae*: physical and genetic characterization. *Molec. Gen. Genet.*, **123**, 51–65.

MICHAELIS, P. (1933). Entwicklungsgeschichtlich-genetische Untersuchungen an *Epilobium*. II. Die Bedeutung des Plasmas für die Pollenfertilität des *Epilobium-luteum-hirsutum* Bastardes. *Z. Vererbungsl.*, **65**, 1–71, 353–411.

MICHAELIS, P. (1965). Cytoplasmic inheritance in *Epilobium* (a survey). *Nucleus*, **8**, 83–92.

MILLER, J. F., SCHMIDT, J. W. & JOHNSON, V. A. (1974). Inheritance of genes controlling male-fertility restoration in the wheat cultivar Primépi. *Crop Sci.*, **14**, 437–8.

MILLER, R. J. & KOEPPLE, D. E. (1971). Southern corn leaf blight: susceptible and resistant mitochondria. *Science*, **173**, 67–9.

MITCHELL, D. J., HERRING, A. J. & BEVAN, E. A. (1976). The genetic control of Ds-RNA virus-like particles associated with *Saccharomyces cerevisiae* killer yeast. *Heredity*, **37**, 129–34.

MITCHELL, M. B. & MITCHELL, H. K. (1952). A case of maternal inheritance in *Neurospora crassa*. *Proc. Natl. Acad. Sci.*, **38**, 442–9.

MOLLOY, P. L., LINNANE, A. W. & LUKINS, H. B. (1975). Biogenesis of mitochondria: analysis of deletion of mitochondrial antibiotic resistance markers in petite mutants of *Saccharomyces cerevisiae*. *J. Bacteriol.* **20**, 7–18.

MORGAN, T. H. (1926). Genetics and the physiology of development. *Amer. Nat.*, **60**, 489–515.

MOUNOLOU, J. C., JAKOB, H. & SLONIMSKI, P. P. (1966). Mitochondrial DNA from yeast 'petite' mutants: specific changes of buoyant density corresponding to different cytoplasmic mutations. *Biochem. Biophys. Res. Commun.*, **24**, 218–24.

MUNKRES, K. D., GILES, N. H. & CASE, M. E. (1965). Genetic control of *Neurospora* malate dehydrogenase and aspartate amino transferase. *Arch. Biochem. Biophys.*, **109**, 397–403.

MUNN, E. A. (1974). *The Structure of Mitochondria*, Academic Press, London and New York.

NAGLEY, P. & LINNANE, A. W. (1972). Biogenesis of mitochondria. XXI *J. Mol. Biol.*, **66**, 181–93.

NAGLEY, P., MOLLOY, P. L., LUKINS, H. B. & LINNANE, A. W. (1974). Studies on mitochondrial gene purification using petite mutants of yeast. *Biochem. Biophys. Res. Commun.*, **57**, 232–9.

NASS, M. M. K. & NASS, S. (1963). Intramitochondrial fibers with DNA characteristics. *J. Cell Biol.*, **19**, 593–611.

NOMURA, M., TISSIÈRES, A. & LENGYEL, P. (1974). *Ribosomes*, Cold Spring Harbor Laboratory, New York.

NORDSTRÖM, K., INGRAM, L. C. & LUNDBÄCK, A. (1972). Mutations of R factors of *E. coli* causing an increased number of R factor copies per chromosome. *J. Bact.*, **110**, 562–9.

NORDSTRÖM, U. M., ENGBERG, B. & NORDSTRÖM, K. (1974). Competition for DNA polymerase III. between the chromosome and the R factor R1. *Molec. Gen. Genet.*, **135**, 185–90.

NOVICK, R. P. & BOUANCHAUD, D. (1971). Extrachromosomal nature of drug resistance in *Staphylococcus aureus*. *Ann. N.Y. Acad. Sci.*, **182**, 279–94.

NOVICK, R. P., WYMAN, L., BOUANCHAUD, D. & MURPHY, E. (1975). Plasmid life cycles in *Staphylococcus aureus*. In: *Microbiology 1974*, ed. D. Schlessinger. American Society for Microbiology, Washington DC, 115–29.

NOVICK, R. P., CLOWES, R. C., COHEN, S. N., CURTISS III, R., DATTA, N. & FALKOW, S. (1976). Uniform nomenclature for bacterial plasmids: a proposal. *Bact. Rev.*, **40**, 168–89.

OISHI, K. (1971). Spirochaete-mediated abnormal sex-ration (SR) condition in *Drosophila*: a second virus associated with spirochaetes and its use in the study of the SR condition. *Genet. Res.*, **18**, 45–56.

OISHI, K. & POULSON, D. F. (1970). A virus associated with SR-spirochaetes of *Drosophila nebulosa*. *Proc. Natl. Acad. Sci.*, **67**, 1565–72.

ORGEL, L. E. (1963). The maintenance of the accuracy of protein synthesis and its relevance to ageing. *Proc. Natl. Acad. Sci.*, **49**, 517–21.

OSSIPOV, D. V. (1976). The specificity of localization of omega particles, the intranuclear symbiotic bacteria in *Paramecium caudatum*. *Acta Protozool.*, **14**, 43–56.

PAJOT, P., WAMBIER-KLUPPEL, M. L., KOTYLAK, Z. & SLONIMSKI, P. P. (1976). Regulation of cytochrome oxidase formation by mutations in a mitochondrial gene for cytochrome b. In: *Genetics and Biogenesis of Chloroplasts and Mitochondria*, eds. Th. Bücher, W. Neupert, W. Sebald and S. Werner. North Holland, Amsterdam. 443–51.

PERLMAN, D. & ROUND, R. H. (1976). Two origins of replication in composite R plasmid DNA. *Nature*, **259**, 281–84.

PERLMAN, P. S. & BIRKY, W. (1974). Mitochondrial genetics in baker's yeast: a molecular mechanism for recombinational polarity and suppressiveness. *Proc. Natl. Acad. Sci.*, **71**, 4612–6.

PETSCHENKO, B. (1911). *Drepanospira Muelleri* n.g.n.sp. parasite des Paraméciums: contribution à l'étude de la structure des bactéries. *Arch. Protistenk.*, **22**, 252–98.

PITTENGER, T. H. (1956). Synergism of two cytoplasmically inherited mutants in *Neurospora crassa*. *Proc. Natl. Acad. Sci.*, **42**, 747–52.

PLISCHKE, M. E., BORSTEL, R. C. VON, MORTIMER, R. K. & COHN, W. E. (1975). *Handbook of Biochemistry and Molecular Biology*, 3rd Ed., ed. C. D. Fasman. Chemical Publ. Press, Cleveland.

PONTECORVO, G. (1959). *Trends in genetic analysis*, Colombia University Press, New York.

POTTER, S. S., NEWBOLD, J. E., HUTCHISON, C. A. & EDGELL, M. H. (1975). Specific cleavage analysis of mammalian mitochondrial DNA. *Proc. Natl. Acad. Sci.*, **72**, 4496–4500.

POULSON, D. F. (1963). Cytoplasmic inheritance and hereditary infections in *Drosophila*. In: *Methodology in Basic Genetics*, ed. W. J. Burdette. Holden-Day, San Francisco, 404–24.

POULSON, D. F. (1968). Nature, stability and expression of hereditary SR infections in *Drosophila*. *Proc. 12th Internat. Conf. Genetics*, **2**, 91–2.

PREER, J. R., PREER, L. B. & JURAND, A. (1974). Kappa and other endosymbionts in *Paramecium aurelia*. *Bact. Rev.*, **38**, 113–63.

PRING, D. R., LEVINGS, C. S., III, HU, W. W. L. & TIMOTHY, D. H. (1977). Unique DNA associated with mitochondria in the 'S'-type cytoplasm of male sterile maize. *Proc. Natl. Acad. Sci.* **74**, 2904–8.

PRINTZ, P. (1973). Relationship of sigma virus to vesicular stomatitis virus. *Adv. Virus Res.*, **18**, 143–57.

PRITCHARD, R. H., BARTH, P. T. & COLLINS, J. (1969). Control of DNA synthesis in bacteria. In: *Microbial Growth*, eds. P. M. Meadow and S. J. Pirt. *Symp. Soc. Gen. Microbiol.* **19**, 263–7. Cambridge University Press, London, 263–7.

PUHALLA, J. E. (1968). Compatibility reactions on solid medium and interstrain inhibition in *Ustilago maydis*. *Genetics*, **60**, 461–74.

PUHALLA, J. E. & SRB, A. M. (1967). Heterokaryon studies of the cytoplasmic mutant SG in *Neurospora*. *Genet. Res.*, **10**, 185–94.

RAFF, R. A. & MAHLER, H. R. (1972). The non-symbiotic origin of mitochondria. *Science*, **177**, 575–82.

RAFF, R. A. & MAHLER, H. R. (1975). The symbiont that never was: an inquiry into the evolutionary origin of the mitochondrion. In: *Symbiosis*, eds. D. H. Jennings and D. L. Lee. *Symp. Soc. Exp. Biol.*, **29**, Cambridge University Press, London, 41–92.

RANK, G. H. & BECH-HANSEN, N. T. (1972). Somatic segregation, recombination, asymmetrical distribution and complementation tests of cytoplasmically-inherited antibiotic-resistance mitochondrial markers in *S. cerevisiae*. *Genetics*, **72**, 1–15.

REEVE, R. & WILLETTS, N. (1974). Plasmid specificity of the origin of transfer of sex factor F. *J. Bact.*, **120**, 125–30.

REIJNDERS, L. (1975). The origin of mitochondria. *J. Mol. Evol.*, **5**, 167–76.

REIJNDERS, L. & BORST, P. (1972). The number of 4-S RNA genes on yeast mitochondrial DNA. *Biochem. Biophys. Res. Comm.*, **47**, 126–33.

RENNER, O. (1937). Zur Kenntnis der Plastiden und Plasmavererbung. *Cytologia (Fujii-Jub. Vol.)*, 644 –53.

RHOADES, M. (1933). The cytoplasmic inheritance of male sterility in *Zea mays*. *J. Genet.*, **27**, 71–95.

RHOADES, M. (1950). Gene induced mutation of a heritable cytoplasmic factor producing male sterility in maize. *Proc. Natl. Acad. Sci.*, **36**, 634–5.

RICHARD-MOLARD, C. (1975). Isolement de lignées cellulaires de *Drosophila melanogaster* de différents genotypes et études de la multiplication de deux variants du rhabdovirus sigma dans ces lignées. *Arch. Virology*, **47**, 139–46.

RICHMOND, M. H. (1968). The plasmids of *Staphylococcus aureus* and their relation to other extrachromosomal elements in bacteria. In: *Advances in Microbial Physiology*, **2**, eds. A. H. Rose and J. F. Wilkinson. Academic Press, London, 43–88.

RIFKIN, M. R. & LUCK, D. J. L. (1971). Defective production of mitochondrial ribosomes in the *Poky* mutant of *Neurospora crassa*. *Proc. Natl. Acad. Sci.*, **68**, 287–90.

RIZET, G. (1952). Les phénomènes de barrage chez *Podospora anserina*. I. Analyse génétique des barrages entre souches S et s. *Rev. Cytol. Biol. Veg.*, **13**, 51–92.

ROBBERSON, D. L., KASAMATSU, H. & VINOGRAD, J. (1972). Replication of mitochondrial DNA. Circular replicative intermediate in mouse L cells. *Proc. Natl. Acad. Sci.*, **69**, 737–41.

ROPER, J. A. (1958). Nucleo-cytoplasmic interactions in *Aspergillus nidulans*. *Cold Spring Harbor Symp.*, **23**, 141–54.

ROTH, T. F. & HELINSKI, D. R. (1967). Evidence for circular DNA forms of a bacterial plasmid. *Proc. Natl. Acad. Sci.*, **58**, 650–7.

ROWLANDS, P. T. & TURNER, G. (1975). Three-marker extranuclear mitochondrial crosses in *Aspergillus nidulans*. *Molec. Gen. Genet.*, **141**, 69–79.

ROWND, R., NAKAYA, R. & NAKAMURA, A. (1966). Molecular nature of the drug resistance factors of the Enterobacteriaceae. *J. Mol. Biol.*, **17**, 376–93.

ROWND, R. H., PERLMAN, D. & GOTO, N. (1975). Structure and replication of R-factor DNA in *Proteus mirabilis*. In: *Microbiology 1974*, ed. D. Schlessinger. American Society for Microbiology, Washington DC, 76–94.

RYAN, R. S., CHIANG, K-S. & SWIFT, H. (1974). Circular DNA from mitochondria of *Chlamydomonas reinhardtii*. *J. Cell Biol.*, **63**, 293a.

SACCONE, C. & KROON, A. M. (1976). *The Genetic Function of Mitochondrial DNA*, Elsevier/North Holland, Amsterdam, 1–354.

SAGE, G. C. M. (1972). The inheritance of fertility restoration in male-sterile wheat carrying cytoplasm derived from *Triticum timopheevi*. *Theor. Appl. Genet.*, **42**, 233–43.

SAGER, R. (1972). *Cytoplasmic Genes and Organelles*, Academic Press, New York and London.

SAGER, R. & LANE, D. (1972). Molecular basis of maternal inheritance. *Proc. Natl. Acad. Sci.*, **69**, 2410–3.

SAGER, R. & RAMANIS, Z. (1973). The mechanism of maternal inheritance in *Chlamydomonas*: biochemical and genetic studies. *Theor. Appl. Genet.*, **43**, 101–8.

SAGER, R. & RAMANIS, Z. (1976). Chloroplast genetics of *Chlamydomonas* I, II & III. *Genetics*, **83**, 303–54.

SAINSARD, A. (1975). Mitochondrial suppressor of a nuclear gene in *Paramecium*.

Nature, **257**, 312–14.
SAINSARD, A. (1976). Gene controlled selection of mitochondria in *Paramecium*. *Molec. Gen. Genet.*, **145**, 23–30.
SANDERS, J. P. M., HEYTING, C., DI FRANCO, A., BORST, P. & SLONIMSKI, P. P. (1976). The organization of genes in yeast mitochondrial DNA. In: *The Genetic Function of Mitochondrial DNA*, eds. C. Saccone and A. M. Kroon. North Holland, Amsterdam, 259–72.
SCHATZ, G. (1975). The biosynthesis of cytochrome c oxidase in baker's yeast. In: *Molecular biology of nucleocytoplasmic relationships*, ed. S. Puiseux-Dao. Elsevier, Amsterdam and New York, 157–70.
SCHATZ, G. & MASON, T. L. (1974). The biosynthesis of mitochondrial proteins. *Ann. Rev. Biochem.*, **43**, 51–87.
SCHIMMER, O. & ARNOLD, C. G. (1970). Hin-und Rücksegregation eines ausser-karyotischen Gens bei *Chlamydomonas reinhardi*. *Molec. Gen. Genet.*, **108**, 33–40.
SCHÖTZ, F. (1974). Untersuchungen über die Plastidenkonkurrenz bei *Oenothera*. IV & V. *Biol. Zentr.*, **93**, 41–64; **94**, 17–25.
SCHWARTZBACH, S. D., HECKER, L. I. & BARNETT, W. E. (1976). Transcriptional origin of *Euglena* chloroplast tRNAs. *Proc. Natl. Acad. Sci.*, **73**, 1984–8.
SCHWEYEN, R. J., WEISS-BRUMMER, B., BACKHAUS, B. & KAUDERWITZ, F. (1976). Localization of seven gene loci on a circular map of the mitochondrial genome of *Saccharomyces cerevisiae*. See Saccone & Kroon (1976), 251–58.
SEECOF, R. (1968). The sigma virus infection of *Drosophila melanogaster*. *Current Topics in Microbiology and Immunology*, **42**, 59–93.
SHANNON, C., ENNS, R., WHEELIS, L., BURCHIEL, K. & CRIDDLE, R. (1973). Alterations in mitochondrial ATP activity resulting from mutation of mitochondrial DNA. *J. Biol. Chem.*, **248**, 3004–11.
SHARP, F. A., HSU, M. T., OHTSUBO, E. & DAVIDSON, N. (1972). Electron microscope heteroduplex studies of sequence relations among plasmids of *E. coli*. I. Structure of F-prime factors. *J. Mol. Biol.*, **71**, 471–97.
SHERMAN, F. (1963). Respiration deficient mutants of yeast. I. *Genetics*, **48**, 375–85.
SHERMAN, F. & STEWARD, J. W. (1971). Genetics and biosynthesis of cytochrome c. *Ann. Rev. Genet.*, **5**, 257–96.
SHOWS, T. B., CHAPMAN, V. M. & RUDDLE, F. H. (1970). Mitochondrial malate dehydrogenase and malic enzyme: Mendelian inherited electrophoretic variants in the mouse. *Biochem. Genet.*, **4**, 707–18.
SIMPSON, L. (1972). The kinetoplast of the hemoflagellates. *Internat. Rev. Cytol.*, **32**, 139–207.
SLONIMSKI, P. & TZAGOLOFF, A. (1976). Localization in yeast mitochondrial DNA of mutations expressed in a deficiency of cytochrome oxidase and for coenzyme QH_2-cytochome c reductase. *Eur. J. Biochem.*, **61**, 27–41.
SMITH, D. R., HOOKER, A. L., LIM, S. M. & BECKETT, J. B. (1971). Disease reaction of thirty sources of male sterile corn to *Helminthosporium maydis* race T. *Crop Sci.*, **11**, 772–3.
SMITH, H. H. (1968). Recent cytogenetic studies in the genus *Nicotiana*. *Adv. Genet.*, **14**, 1–54.
SMITH, J. R. & RUBENSTEIN, I. (1973a). The development of 'senescence' in *Podospora anserina*. *J. Gen. Microbiol.*, **76**, 283–94.
SMITH, J. R. & RUBINSTEIN, I. (1973b). Cytoplasmic inheritance of the timing of 'senescence' in *Podospora anserina*. *J. Gen. Microbiol.* **76**, 297–304.
SMITH, S. M. & STOCKER, B. A. D. (1966). Mapping prophage P22 in *Salmonella typhimurium*. *Virology*, **28**, 413–9.
SOMERS, J. M. & BEVAN, E. A. (1969). The inheritance of the killer character in yeast. *Genet. Res. Camb.*, **13**, 71–83.
SONNEBORN, T. M. (1938). Mating types in *P. aurelia*: diverse conditions for mating in different stocks; occurrence, number and inter-relations of the types. *Proc. Amer.*

Phil. Soc., **79**, 411–34.

SONNEBORN, T. M. (1943). Gene and cytoplasm. I. The determination and inheritance of the killer character in variety 4 of *P. aurelia*. II. The bearing of determination and inheritance of characters in *P. aurelia* on problems of cytoplasmic inheritance, pneumococcus transformations, mutations and development. *Proc. Natl. Acad. Sci.*, **29**, 329–43.

SONNEBORN, T. M. (1950). Methods in the general biology and genetics of *Paramecium aurelia*. *J. Exp. Zool.*, **113**, 87–148.

SONNEBORN, T. M. (1959). Kappa and related particles in Paramecium. *Adv. Virus Res.*, **6**, 231–356.

SONNEBORN, T. M. (1974). *Paramecium aurelia*. In: *Handbook of Genetics, Vol. 2.* ed. R. C. King. Plenum Press, New York, 469–594.

SONNEBORN, T. M. (1975). The *Paramecium aurelia* complex of fourteen sibling species. *Trans. Amer. Micros. Soc.*, **94**, 155–78.

SPENCER, R. & CROSS, G. A. M. (1975). Purification and properties of nucleic acids from an unusual cytoplasmic organelle in the flagellate protozoan *Crithidia oncopelti*. *Biochim. Biophys. Acta*, **390**, 141–54.

SPOLSKY, C. M. & EISENSTADT, J. M. (1972). Chloramphenicol-resistant mutants of human HeLa cells. *FEBS Letters*, **25**, 319–24.

SRB, A. M. (1963). Extrachromosomal factors in the genetic differentiation of Neurospora. *Symp. Soc. Exp. Biol.*, **17**, 175–87.

SRB, A. M. (1966). Extrachromosomal heredity in fungi. In: *Reproduction : Molecular, Subcellular and Cellular*, ed. M. Locke. *Symp. Soc. Devel. Biol.*, **24**, Academic Press, New York, 191–211.

SRB, A. M. (1965). Extrachromosomal heredity in fungi. In: *Reproduction : Molecular, Subcellular and Cellular*, ed. M. Locke. *Symp. Soc. Dev. Biol.*, **24**, Academic Press, New York, 191–211.

STANIER, R. Y. (1970). Some aspects of the biology of cells and their possible evolutionary significance. In: *Organization and Control in Prokaryotic and Eukaryotic Cells*, eds. H. P. Charles and B. C. J. G. Knight. *Symp. Soc. Gen. Microbiol.*, **20**, Cambridge University Press, London, 1–38.

STANIER, R. Y. & van NIEL, C. B. (1962). The concept of a bacterium. *Arch. Mikrobiol.*, **42**, 17–35.

STEINERT, M. & van ASSEL, S. (1975). Large circular mitochondrial DNA in *Crithidia luciliae*. *Exp. Cell Res.*, **96**, 406–9.

STENT, G. (1971). *Molecular Genetics: An Introductory Narrative*, Freeman, San Francisco.

STEVENS, B. (1977). Variation in number and volume of the mitochondria in yeast according to growth conditions. A study based on computer graphics reconstitution. *Biol. Cellulaire*, **28**, 37–56.

STUBBE, W. (1959). Genetische Analyse des Zussamenwirkens von Genom und Plastom bei *Oenothera*. *Z. Vererbungsl.*, **90**, 288–98.

STUBBE, W. (1964). The role of the plastome in evolution of the genus *Oenothera*. *Genetica*, **35**, 28–33.

STUTZ, E. (1971). Characterization of *Euglena gracilis* chloroplast single strand DNA. In: *Autonomy and Biogenesis of Mitochondria and Chloroplasts*, eds. N. K. Boardman, A. W. Linnane and R. M. Smillie. North Holland, Amsterdam.

SWEENEY, T. K., TATE, A. & FINK, G. R. (1976). A study of the transmission and structure of double stranded RNAs associated with the killer phenomenon in *Saccharomyces cerevisiae*. *Genetics*, **84**, 27–42.

SZOLLOSI, D. (1965). The fate of sperm middle-piece mitochondria in the rat egg. *J. Exp. Zool.*, **159**, 367–78.

TAIT, A. (1968). Genetic control of β-hydroxybutyrate dehydrogenase in *Paramecium aurelia*. *Nature*, **219**, 941.

TAIT, A. (1970). Genetics of NADP isocitrate dehydrogenase in *Paramecium aurelia*. *Nature*, **225**, 181–2.

TAIT, A. (1972). Altered mitochondrial ribosomes in an erythromycin resistant mutant of *Paramecium. FEBS Letters*, **24**, 117–20.

TAIT, A., & KNOWLES, J. K. C. (1977). Characterization of mitochondrial and cytoplasmic ribosomes from *Paramecium aurelia. J. Cell Biol.*, **73**, 139–48.

TAIT, A., KNOWLES, J. K. C. & HARDY, J. C. (1976a). The genetic control of mitochondrial ribosomal proteins in *Paramecium*. In: *The Genetic Function of Mitochondrial DNA*, eds. C. Saccone and A. M. Kroon. Elsevier/North Holland, Amsterdam 131–6.

TAIT, A., KNOWLES, J. K. C., HARDY, J. C. & LIPPS, H. (1976b). The study of the genetic function of *Paramecium* mitochondrial DNA using species hybrids. In: *The Genetics and Biogenesis of Chloroplasts and Mitochondria*, eds. T. Bücher, W. Neupert, W. Sebald and S. Werner. North Holland, Amsterdam, 569–72.

TAYLOR, F. J. R. (1974). Implications and extensions of the serial endosymbiosis theory of the origin of eukaryotes. *Taxon*, **23**, 229–58.

TEWARI, K. K. & WILDMAN, S. G. (1970). Information content in the chloroplast DNA. In: *Control of Organelle Development*, ed. P. L. Miller. *Symp. Soc. Exp. Biol.*, **24**, Cambridge University Press, London, 147–80.

THOMAS, D. Y. & WILKIE, D. (1968a). Recombination of mitochondrial drug-resistance factors in *Saccharomyces cerevisiae. Biochem. Biophys. Res. Commun.*, **30**, 368–72.

THOMAS, D. Y. & WILKIE, D. (1968b). Inhibition of mitochondrial synthesis in yeast by erythromycin: cytoplasmic and nuclear factors controlling resistance. *Genet. Res.*, **11**, 33–41.

TILNEY-BASSETT, R. A. E. (1970). The control of plastid inheritance in *Pelargonium. Genet. Res.*, **16**, 49–61.

TREMBATH, M. K., MOLLOY, P. L., SRIPRAKASH, K. S., CUTTING, G. S., LINNANE, A. W. & LUKINS, H. B. (1976). Biogenesis of Mitochondria 44. Comparative studies and mapping of mitochondrial oligomycin resistance mutants in yeast based on gene recombination and petite deletion analysis. *Molec. Gen. Genet.*, **145**, 43–52.

TUVESON, R. W. & PATERSON, J. F. (1972). Virus-like particles in certain slow-growing strains of *Neurospora crassa. Virology*, **47**, 527–31.

TZAGOLOFF, A. (ed.) (1975). *Membrane Biogenesis: Mitochondria Chloroplasts and Bacteria*, Plenum, New York and London.

TZAGOLOFF, A., AKAI, A., NEEDLEMAN, R. B. & ZULCH, G. (1975). Assembly of the mitochondrial membrane system. Cytoplasmic mutants of *Saccharomyces cerevisiae* with lesions in enzymes of the respiratory chain and in the mitochondrial ATPase. *J. Biol. Chem.*, **250**, 8236–42.

TZAGOLOFF, A., FOUVY, F. & AKAI, A. (1976). Genetic loci on mitochondrial DNA involved in cytochrome b biosynthesis. *Molec. Gen. Genet.*, **149**, 33–42.

UPHOLT, B. U. & BORST, P. (1974). Accumulation of replicative intermediates of mitochondrial DNA in *Tetrahymena pyriformis* grown in ethidium bromide. *J. Cell Biol.*, **61**, 383–97.

URSPRUNG, H. & SCHABTACH, E. (1965). Fertilization in Tunicates. Loss of the paternal mitochondrion prior to sperm entry. *J. Exp. Zool.*, **159**, 379–84.

UZZEL, T. & SPOLSKY, C. (1974). Mitochondria and plastids as endosymbionts: a revival of special creation? *Amer. Sci.*, **62**, 334–43.

VICKERMAN, K. (1965). Polymorphism and mitochondrial activity in sleeping sickness trypanosomes. *Nature*, **208**, 762–6.

VIGNAIS, P. V., HUET, J. & ANDRÉ, J. (1969). Isolation and characterization of ribosomes from yeast mitochondria. *FEBS Letters*, **3**, 177–81.

VISCONTI, N. & DELBRÜCK, M. (1953). The mechanism of genetic recombination in phage. *Genetics*, **38**, 5–33.

WAGTENDONK, W. J. van, CLARK, J. A. D. & GODOY, G. A. (1963). The biological status of lambda and related particles in *P. aurelia. Proc. Natl. Acad. Sci.*, **50**, 835–8.

WATANABE, T. (1963). Infective heredity of multiple drug resistance in bacteria. *Bact. Rev.*, **26**, 23–8.

WATANABE, T. & OGATA, C. (1966). Episome-mediated transfer of drug resistance in *Enterobactericeae*. IX. Recombination of an R factor with F. *J. Bact.*, **91**, 43–50.

WEHRMEYER, W. (1964). Über Membranbildungsprozesse im Chloroplasten. II. Mitteilung: Zur Entstehung der Grana durch Membranüberschiebung. *Planta*, **63**, 13–30.

WESTERGAARD, O., MARCKER, K. A. & KEIDING, J. (1970). Induction of a mitochondrial DNA polymerase in *Tetrahymena*. *Nature*, **227**, 708–10.

WETTSTEIN, D. VON, HENNINGSEN, K. W., BOYNTON, J. E., KANNAGAWA, G. C. & NIELSEN, O. F. (1971). The genetic control of chloroplast development in barley. In: *Anatomy and Biogenesis of Mitochondria and Chloroplasts*, eds. N. K. Boardman, A. W. Linnane and R. M. Smillie. North Holland, Amsterdam, 205–23.

WETTSTEIN, F. VON (1946). Untersuchungen zur plasmatischen Vererbung. I. *Linum*. *Biol. Zentr.*, **65**, 149–66.

WHATLEY, J. M. (1976). Bacteria and nuclei in *Pelomyxa palustris*: comments on the theory of serial endosymbiosis. *New Phytol.*, **76**, 111–20.

WHITEHOUSE, H. L. K. (1968). *Towards an understanding of the mechanism of heredity*, 2nd Ed., Edward Arnold, London.

WICKNER, R. B. & LEIBOWITZ, M. J. (1976). Two chromosomal genes required for killing expression in killer strains of *Saccharomyces cerevisiae*. *Genetics*, **82**, 429–42.

WILDMAN, S. G., HONGLADAROM, T. & HONDA, S. I. (1962). Chloroplasts and mitochondria in living plant cells: cinephotomicrographic studies. *Science*, **138**, 434–6.

WILKIE, D. & THOMAS, D. Y. (1973). Mitochondrial genetic analysis by zygote cell lineages in *Saccharomyces cerevisiae*. *Genetics*, **73**, 367–77.

WILKINS, B. M. & HOLLOM, S. E. (1974). Conjugational synthesis of F *lac*+ and Col. 1 DNA in the presence of rifampicin and *E. coli* K12 mutants defective in DNA synthesis. *Molec. Gen. Genet.*, **134**, 143–56.

WILLETTS, N. S. (1975). The genetics of conjugation. In: *R factors*, ed. S. Mitsuhashi. University of Tokyo Press, Tokyo.

WILLETTS, N. S. & ACHTMAN, M. (1972). A genetic analysis of transfer by the *E. coli* sex factor F, using P1 transductional complementation. *J. Bact.*, **110**, 843–51.

WILLIAMS, J. (1971). The growth *in vitro* of killer particles from *P. aurelia* and the axenic culture of this protozoan. *J. Gen. Microbiol.*, **68**, 253–62.

WILLIAMSON, D. (1961). Carbon dioxide sensitivity in *Drosophila affinis* and *D. athabasco*. *Genetics*, **46**, 1053–60.

WILLIAMSON, D. H. & FENNELL, D. J. (1974). Apparent dispersive replication of yeast mitochondrial DNA as revealed by density labelling experiments. *Molec. Gen. Genet.*, **131**, 193–207.

WILLIAMSON, D. L. (1966). A typical transovarial transmission of sex ratio spirochetes by *Drosophila robusta* Sturtevant. *J. exp. Zool.*, **161**, 425–28.

WILLIAMSON, D. L. & WHITCOMBE, R. F. (1974). Helical, wall free prokaryotes in *Drosophila*, leaf-hoppers and plants. *Colloques de l'Institut National de la Santé et de la Recherche Médicale, Paris*, **33**, 283–90.

WILSON, J. F. (1969). Mitochondrial transplantation studies in *Neurospora crassa* (Abstract). *Proc. XI Int. Bot. Cong.* Seattle, Washington, 240.

WINTERSBERGER, E. (1970). DNA-dependent RNA polymerase from mitochondria of a cytoplasmic 'petite' mutant of yeast. *Biochem. Biophys. Res. Commun.*, **40**, 1179–84.

WISEMAN, A., GILLHAM, N. W. & BOYNTON, J. E. (1977). The mitochondrial genome of *Chlamydomonas*. II Genetic analysis of non-Mendelian obligate photoautotrophic mutants. *Molec. Gen. Genet.* **150**, 109–18.

WOLF, K., DUJON, B. & SLONIMSKI, P. P. (1973). Mitochondrial genetics, V. Multifactorial mitochondrial crosses involving a mutation conferring paromomycin-resistance in *Saccharomyces cerevisiae*. *Molec. Gen. Genet.*, **125**, 53–90.

WOOD, H. A. & BOZARTH, R. F. (1973). Heterokaryon transfer of virus-like particles and a cytoplasmically inherited determinant in *Ustilago maydis*. *Phytopathology*, **63**, 1019–20.

WU, M., DAVIDSON, N., ATTARDI, G. & ALONI, Y. (1972). Expression of the mitochondrial genome in HeLa cells. XIV. *J. Mol. Biol.*, **71**, 81–93.

YEN, J. H. & BARR, A. R. (1971). New hypothesis of the causes of cytoplasmic incompatibility in *Culex pipiens*. *Nature*, **232**, 657–8.

YEN, J. H. & BARR, A. R. (1974). Incompatibility in *Culex pipiens*. In: *The Use of Genetics in Insect Control*, eds. R. Pal and M. J. Whitten. Elsevier and North Holland, Amsterdam, 97–118.

YOTSUYANAGI, Y. (1962*a*). Études sur le chondriome de la levure. I. Variation de l'ultrastructure du chondriome au course du cycle de la croissance aerobique. *J. Ultrastruct. Res.*, **7**, 121–40.

YOTSUYANAGI, Y. (1962*b*). Études sur le chondriome de la levure. II. Chondriomes des mutants à deficience respiratoire. *J. Ultrastruct. Res.*, **7**, 141–58.

YOUNG, C. S. H. & COX, B. S. (1972). Extrachromosomal elements in a super-suppression system of yeast. *Heredity*, **28**, 189–99.

YUDIN, A. L. (1973). Nuclear–nuclear interactions in heterokaryons. In: *The Biology of Amoeba*, ed. K. W. Jeon. Academic Press, New York and London, 505–23.

ZABLEN, L. B., KISSIL, M. S., WOESE, C. R. & BUETOW, D. E. (1975). Phylogenetic origin of the chloroplast and prokaryotic nature of its ribosomal RNA. *Proc. Natl. Acad. Sci.*, **72**, 2418–22.

Index